Ergebnisse der Mathematik und ihrer Grenzgebiete 99

A Series of Modern Surveys in Mathematics

Bruce L. Reinhart

Differential Geometry of Foliations

The Fundamental Integrability Problem

Springer-Verlag
Berlin Heidelberg New York Tokyo
1983

Bruce L. Reinhart
University of Maryland
Department of Mathematics
College Park, MD 20742, USA

AMS Subject Classification (1980): 53C15

Library of Congress Cataloging in Publication Data. Reinhart, Bruce L. Differential geometry of foliations. (Ergebnisse der Mathematik und ihrer Grenzgebiete; 99). Includes bibliographies and index. 1. Foliations (Mathematics). 2. Geometry, Differential. I. Title. II. Series. QA613.62.R44. 1983. 514'.72. 83-4711.

ISBN-13: 978-3-642-69017-4 e-ISBN-13: 978-3-642-69015-0
DOI: 10.1007/978-3-642-69015-0

Softcover reprint of the hardcover 1st edition 1983

Typesetting: Daten- und Lichtsatz-Service, Würzburg

2141/3140-543210

Preface

> Whoever you are! How can I but offer you divine leaves...?
>
> Walt Whitman

The object of study in modern differential geometry is a manifold with a differential structure, and usually some additional structure as well. Thus, one is given a topological space M and a family of homeomorphisms, called coordinate systems, between open subsets of the space and open subsets of a real vector space V. It is supposed that where two domains overlap, the images are related by a diffeomorphism, called a coordinate transformation, between open subsets of V. M has associated with it a tangent bundle, which is a vector bundle with fiber V and group the general linear group $GL(V)$. The additional structures that occur include Riemannian metrics, connections, complex structures, foliations, and many more. Frequently there is associated to the structure a reduction of the group of the tangent bundle to some subgroup G of $GL(V)$. It is particularly pleasant if one can choose the coordinate systems so that the Jacobian matrices of the coordinate transformations belong to G. A reduction to G is called a G-structure, which is called integrable (or flat) if the condition on the Jacobians is satisfied. The strength of the integrability hypothesis is well-illustrated by the case of the orthogonal group O_n. An O_n-structure is given by the choice of a Riemannian metric, and therefore exists on every smooth manifold. On the other hand, it is integrable only if the curvature of every 2-plane is 0, a very severe restriction. Given a differential manifold, it is natural to inquire whether it admits a G-structure, and if so whether an effective classification of such structures can be given. Many of the methods of modern algebraic topology were developed to answer these questions, and in fact they have yielded a great deal of information on the answers. The corresponding questions for integrable G-structures require in addition analytic methods. To see why this is so, consider the problem of constructing a complex structure on a 2-dimensional manifold, that is, an integrable reduction of the group of the tangent bundle from $GL(\mathbb{R}^2)$ to the subgroup $GL(\mathbb{C})$. The latter is isomorphically embedded in the former by the mapping

$$(\alpha + i\beta) \mapsto \begin{pmatrix} \alpha & \beta \\ -\beta & \alpha \end{pmatrix}.$$

If a coordinate transformation is of the form

$$(x, y) \mapsto (u, v)$$

then the conditions on its Jacobian matrix can be written

$$\frac{\partial u}{\partial x} = \frac{\partial v}{\partial y}, \quad \frac{\partial u}{\partial y} = -\frac{\partial v}{\partial x}.$$

Thus, to study the existence of a complex structure, one must study the Cauchy-Riemann equations. In general, the study of an integrable G-structure requires the study of a system of partial differential equations. These equations are of very different types for different choices of G, so that the properties of a flat G-structure vary radically as G varies. Nonetheless, since the given object is a differential manifold with no further structure, the existence and classification problems should be solved in terms of invariants defined without the use of further structures. The invariants now known to differential topologists are of this character, but in the best known cases, these provide countable classifications, and are therefore insufficient. It is to be hoped that global analysis will provide the additional invariants needed. In particular, there are a number of results in the theory of foliations that seem to be capable of generalization.

It is impossible to describe here in any detail the steps which brought differential geometry to its present form. Indeed, the problem of determining the tangent lines to a curve antedates by several millenia the invention of the derivative, which it helped to motivate. The studies of perspective by the artists of the Renaissance developed into projective geometry, in which complex numbers came eventually to play an important role. The introduction of coordinate geometry was essential, not only in the recognition of the role of complex numbers, but also in increasing the supply both of configurations available for study and of methods for studying them. The study of curves and surfaces in 3-dimensional space combined with the discovery of noneuclidean geometries to motivate the introduction of Riemannian geometry. Symplectic geometry grew out of mechanics. But the recognition of the common structure underlying all these tendencies was late in coming, and there may still be fundamental discoveries to be made.

For the last hundred years, geometry has been viewed primarily as the study of properties invariant under coordinate changes. If the coordinate changes form a group, there results a beautiful theory full of deep results, and having a strongly algebraic flavor. If the group is a real or complex Lie group, the geometry has differential aspects. Unfortunately, the coordinate changes in differential geometry do not usually form a group, but some more general structures. There are various methods by which differential geometry has been brought at least partially into the general trend. One is to consider the tangent space at each point as an object in its own right. It is a vector space, so has a distinguished element 0 and a natural group, the general linear group. At each point, therefore, one has invariants associated with this group and its subgroups, acting not only on the tangent space but on various other spaces, such as tensor spaces, constructed from it. But it is necessary to consider how these vary from point to point. First of all, the subgroup must vary smoothly, so that the collection of all subgroups forms a principal bundle over the manifold. These bundles map contravariantly, and they are classified up to a suitable equivalence relation by homotopy classes of maps into a classifying space. The cohomology ring of the classifying space, pulled back by the classifying map, becomes the characteristic ring of the bundle. The general theory works for any topological group (and even more generally), and in the case of the best known Lie groups, there are available powerful methods of calculating the characteristic classes. A connection is introduced to make it possible to slide invariant objects along a curve and thereby compare them from point to point. But treating the tangent space as a vector space implies that there is a distinguished element, zero, which is geometrically very unnatural, since all points of a manifold are equivalent under its diffeomorphism group. Thus, it is better to treat the tangent space as an affine space, that is, the group is extended from a subgroup of the general linear group to the corresponding subgroup of the affine group by putting the translations into it. In the corresponding principal bundle, the additional components of the curvature thus obtained may be identified with the torsion of the connection. But the resulting theory still does not capture the essence of integrability, since the curvature of the (linear or affine) connection usually is subjected by the integrability assumption to certain restrictions which are weaker than integrability. In some cases, the situation is improved by taking explicit account of derivatives up to some higher, but finite, order. The general linear group is replaced by a group consisting of mappings, whose components are polynomials with 0 constant term and order at most k, where terms of order greater than k in a composition are discarded. But this still does not in general provide a complete answer, at least at present, since very little is known about these groups and their representations. Furthermore, the extension to the affine group does not even exist, since allowing nonzero constant terms makes the k-th order terms of a composition depend upon higher order terms of the factors. Thus, one is led to try additional approaches.

A coordinate change is a local diffeomorphism, that is, it is defined on some open subset. The set of local diffeomorphisms forms a pseudogroup. The concept of pseudogroup is hard to work with, but some results have been obtained. Further results are obtained by considering germs of diffeomorphisms. (Two diffeomorphisms determine the same germ at a point if they agree on some neighborhood of the point.) These form a groupoid, in the sense already known in the theory of quadratic forms, and have a natural topology consistent with their algebraic structure. A topological groupoid has a classifying space, and the characteristic ring which arises from it does take account of integrability. Unfortunately, the object classified is some sort of singular structure, definable on an arbitrary topological space, which even on a smooth manifold may not be equivalent to a smooth nonsingular object of the type one wants. However, in the case of foliations, the bundle and groupoid approaches together provide a global solution of the classification problem. For most other differential geometric structures, this problem still awaits solution. Indeed, effective determination even of local equivalence is not always possible.

Lie algebras play an important part in the theory of Lie groups, so it is useful to notice that there are Lie algebras associated to the pseudogroups and groupoids of interest in differential geometry. Indeed, by the local integrability theorem, any subalgebra of the Lie algebra of C^∞ vector fields on a manifold generates a pseudogroup of C^∞ diffeomorphisms. Given a point and a coordinate system in the neighborhood of the point, the Taylor series at the point of the components of a C^∞ vector field combine to give a formal vector field

$$\sum_{r=0}^{\infty} \sum_{i,\, i_\alpha=1}^{m} C^i_{i_1 \dots i_r} x^{i_1} \dots x^{i_r} \frac{\partial}{\partial x^i}$$

where $C^i_{i_1 \ldots i_r}$ is a real constant. The formal vector fields arising from a Lie algebra of vector fields also make up a Lie algebra, which is more properly associated to the corresponding groupoid, since it depends only on germs at the point. As linear topological spaces, these Lie algebras are nuclear, as are most of the spaces which occur in C^∞ differential geometry. Nuclear spaces have the very useful property that for them, tensor algebra behaves as it does for finite dimensional spaces. On the other hand, though there is a good implicit function theorem valid for many of the interesting C^∞ spaces, it is known that it cannot be extended to all nuclear spaces, or even to all the important spaces of C^∞ objects on a noncompact manifold. Thus, there is at the moment no general theory that unifies the analytic study of C^∞ manifolds. If such a theory could be established, it would subsume the C^k theory. Indeed, each C^k manifold for $k \geqslant 1$ is C^k diffeomorphic to a C^∞ manifold, and the usual function spaces involving finitely many derivatives can be obtained by completing some space of C^∞ objects with respect to a suitable topology.

It is time to see how foliations, which are the principal object of study in this book, fit into the general context of structures defined by coordinate changes. Indeed, they are essential, since the natural notion of quotient in this context is precisely that of foliation. Before this remark can be justified, some discussion of foliations must be given. To describe the coordinate changes on a foliated manifold, consider linear coordinates

$$(x^1, \ldots, x^p, y^1, \ldots, y^q)$$

in $\mathbb{R}^{p+q}$, and consider the submersion of $\mathbb{R}^{p+q}$ to $\mathbb{R}^q$ obtained by mapping (x, y) to (y). The coordinate changes on a foliated manifold must commute with this projection, hence must be of the form

$$(x, y) \mapsto (f(x, y), g(y)).$$

Thus, the group of the tangent bundle can be reduced to matrices of the form

$$\begin{pmatrix} A & B \\ 0 & D \end{pmatrix}.$$

Furthermore, the p-dimensional manifolds defined locally by setting y equal to a constant join together to form immersed submanifolds, not necessarily closed, called leaves. Pictorially, a foliation is a partition of a manifold which looks locally like a family of parallel planes in euclidean space. For example, a locally trivial fiber space is foliated by the fibers, and a product space has a pair of transversal foliations. Some more interesting examples are the foliation whose leaves are the integral curves of a nonsingular vector field and the one whose leaves are the inverse images of a point under a submersion. These examples give only a dim idea of the complicated geometric possibilities that any general theory must apply to.

A foliation may equally well be defined by a family of submersions defined on open subsets to $\mathbb{R}^q$. Indeed, given a foliation defined by coordinate systems, the composition of the map to $\mathbb{R}^{p+q}$ defined by a coordinate system and the projection from $\mathbb{R}^{p+q}$ to $\mathbb{R}^q$ is a submersion, and two such are related by the y-component $g(y)$ of the coordinate change. Conversely, coordinate systems can be constructed from a family of submersions suitably related by local diffeomorphisms of $\mathbb{R}^q$. However, local submersion and local immersion are the natural generalizations for manifolds of unequal dimension of the notion of local diffeomorphism. It is in this sense that foliations are the natural quotient objects in differential geometry. (The local immersions defining individual leaves do not carry all the information, since they do not describe the relations among different leaves. Indeed, even for a product space, it is the projections one must work with.) In general, the global quotient obtained by identifying each leaf to a point is uninteresting, since its natural topology is too coarse, and one studies foliations by means of various constructions associated to local quotients and appropriately related by mappings associated to the family of local diffeomorphisms $g(y)$.

Another point of view is that a foliation is a subbundle of the tangent bundle such that the space of sections is a Lie algebra. From this point of view, many ideas from topological dynamics can be seen to be useful in foliation theory.

The preceding paragraphs are intended to sketch the main ideas which motivate this presentation of foliation theory and give it unity. The first chapter may be viewed as a more detailed introduction. Further historical details and citations can be found throughout the book, usually at the end of sections. Finally, the chapter introductions serve to put various aspects of the theory into the proper perspective.

March, 1983 Bruce L. Reinhart

Table of Contents

Chapter I. Differential Geometric Structures and Integrability

The purpose of this chapter is to introduce the notion of foliation as a particular and fundamental example of a differential geometric structure. Thus, the first step is to introduce various approaches to the definition of a structure, and show how the principal examples, especially foliations, fit in. Next, since a foliation can be viewed as an integrable reduction of the group of the tangent bandle, some general facts about integrability are introduced. As an illustration, the case where the coordinate changes are translations in euclidean space is worked out in detail. Then other important examples, especially foliations, are discussed. In this discussion, the case of only finitely many derivatives is considered carefully, since the difference among orders of differentiability is of increasing geometric significance. Finally, some concepts of particular usefulness in foliation theory are studied and a variety of examples given as motivation for later chapters.

1. Pseudogroups and Groupoids

A manifold is built out of open sets of euclidean space, glued together by homeomorphisms. Various structures on the manifold are obtained by restricting the choice of homeomorphisms. In this section, the construction of a structured manifold will be discussed in detail and examples given.

One way to formulate the definition is by use of the idea of a pseudogroup. A pseudogroup $\mathscr{G}$ is a collection of homeomorphisms, each with domain and image an open subset of some fixed topological space called the model space. $\mathscr{G}$ satisfies certain conditions which make it possible to construct other spaces by using open subsets of the model space, attached by homeomorphisms belonging to $\mathscr{G}$. In differential geometry, the model space is usually one of the spaces $\mathbb{R}^m = \{(x^1, \ldots, x^m) | x^i \text{ is a real number}\}$, and $\mathscr{G}$ consists of homeomorphisms that preserve some additional structure given on $\mathbb{R}^m$. Any mapping of a subset of $\mathbb{R}^m$ into $\mathbb{R}^m$ can be written in terms of components as

$$f(x^1, \ldots, x^m) = (f^1(x^1, \ldots, x^m), \ldots, f^m(x^1, \ldots, x^m)).$$

For a pseudogroup in $\mathbb{R}^m$, we assume that all components of all elements are of a fixed differentiability class C^k, where $k = 0, 1, 2, \ldots, \infty, \omega$. $f \in C^\omega$ means that the Taylor series for f^i at any point converges to f^i in some neighborhood of that point; otherwise $f \in C^k$ means that each f^i has continuous partial derivatives of all orders up to k.

1.1 Definition. Let X be a topological space. Then a *pseudogroup* $\mathcal{G}$ *of homeomorphisms of* X is a collection $\{f_\mu \mid \mu \in \mathcal{M}\}$ of homeomorphisms with domain U_μ and range V_μ open subsets of X, such that:

a) The identity map of X belongs to $\mathcal{G}$.
b) if $f_\mu \in \mathcal{G}$, then its restriction to any open subset of its domain belongs to $\mathcal{G}$.
c) If $f_\mu \in \mathcal{G}$, then $f_\mu^{-1} \in \mathcal{G}$.
d) If $f_\mu, f_\nu \in \mathcal{G}$, $f_\mu\colon U_\mu \to V_\mu$, $f_\nu\colon U_\nu \to V_\nu$, and $V_\mu \cap U_\nu \neq \phi$, then
$$f_\nu \circ f_\mu\colon f_\mu^{-1}(V_\mu \cap U_\nu) \to f_\nu(V_\mu \cap U_\nu)$$
belongs to $\mathcal{G}$.
e) If $f_\mu \in \mathcal{G}$ for $\mu \in \mathcal{M}_1 \subset \mathcal{M}$, $f_\mu\colon U_\nu \to V_\mu$, and $f_\mu | U_\mu \cap U_\nu = f_\nu | U_\mu \cap U_\nu$ for $(\mu, \nu) \in \mathcal{M}_1 \times \mathcal{M}_1$, and if the mapping $f\colon \bigcup_{\mu \in \mathcal{M}_1} U_\mu \to \bigcup_{\mu \in \mathcal{M}_1} V_\mu$ defined by $f | U_\mu = f_\mu$ is a homeomorphism, then $f \in \mathcal{G}$.

$\mathcal{G}$ is called *transitive* if in addition it satisfies:

f) Given any two points of the model space X, there is an element of $\mathcal{G}$ that takes one to the other.

The properties a) through e) may be summarized by saying that $\mathcal{G}$ contains the identity and is closed under restriction, (the taking of) inverses, composition (when it is defined), and union (when the union is a homeomorphism). If the first four properties are satisfied, then a well-defined pseudogroup is obtained by taking as additional elements all those homeomorphisms which are unions of elements of $\mathcal{G}$. If $x \in X$, the orbit of x is $\{f_\mu(x) \mid \mu \in \mathcal{M}\}$, so $\mathcal{G}$ is transitive if and only if the orbit of each point is X.

The idea of morphism of a pseudogroup must also be defined locally. To do so, it will be convenient to use abbreviated statements of the form just introduced.

1.2 Definition. Let $\mathcal{G}^i$ be a pseudogroup of homeomorphisms of X^i, $i = 1, 2$. A *morphism* $\Phi\colon \mathcal{G}^1 \to \mathcal{G}^2$ is a collection $\{\phi \mid \nu \in \mathcal{N}\}$ of homeomorphisms with domain U_ν^1 open in X^1 and image V_ν^2 open in X^2 such that:

a) $\bigcup_{\nu \in \mathcal{N}} U_\nu^1 = X^1$.
b) Φ is closed under restriction.
c) If $f_\mu^1 \in \mathcal{G}^1$, and $\phi_\lambda, \phi_\nu \in \Phi$, then $\phi_\nu \circ f_\mu^1 \circ \phi_\lambda^{-1} \in \mathcal{G}^2$ whenever it is defined.
d) If $f_\mu^1 \in \mathcal{G}^1$, $f_\lambda^2 \in \mathcal{G}^2$, and $\phi_\nu \in \Phi$, then $f_\lambda^2 \circ \phi_\nu \circ f_\mu^1 \in \Phi$ whenever it is defined.
e) Φ is closed under unions, when the union is a homeomorphism.

If the $\mathcal{G}^1$-orbit of each point of X^1 meets the domain of some $\phi \in \Phi$ and the properties b) and c) hold, then Φ can be uniquely completed to a morphism by adjoining all unions of elements of the form $f_\lambda^2 \circ \phi_\nu \circ f_\mu^1$, provided $f_\mu^1 \in \mathcal{G}^1$, $f_\lambda^2 \in \mathcal{G}^2$, $\phi_\nu \in \Phi$, and the union is a homeomorphism. In particular, the composition of morphisms is (contained in) a morphism. Any pseudogroup $\mathcal{G}$ can be viewed as a morphism $\mathcal{G}\colon \mathcal{G} \to \mathcal{G}$, and when so viewed is an identity for the composition of morphisms, in the sense of Definition 1.3 below. Moreover, the morphism $\Phi\colon \mathcal{G}^1 \to \mathcal{G}^2$ is an isomorphism if and only if there is a morphism $\Psi\colon \mathcal{G}^2 \to \mathcal{G}^1$ such that $\Psi \circ \Phi = \mathcal{G}^1$ and $\Phi \circ \Psi = \mathcal{G}^2$.

1.3 Definition. If Γ is a collection such that a composition is defined between certain pairs of elements, an *identity* is an element ε of Γ such that whenever $\alpha\varepsilon$ is defined, $\alpha\varepsilon = \alpha$, and whenever $\varepsilon\alpha$ is defined, $\varepsilon\alpha = \alpha$.

In defining the idea of structure on a manifold, the fundamental concept is that of pseudogroup. However, the collection of germs of elements of a pseudogroup also forms an interesting object called a groupoid. The precise relationship will be discussed after the necessary definitions have been made.

1.4 Definition. Given a topological space X, a *map germ* at a point x of X is an equivalence class of maps, each defined on some neighborhood of x, such that two maps are equivalent if their restrictions to some neighborhood of x are equal. Given a map f, the germ of f at x is denoted by f_x.

Note that f_x can be evaluated at x, but at no other point.

1.5 Definition. A *groupoid* is a set Γ together with a law of composition defined for some pairs of element such that:

a) If either $(\alpha\beta)\gamma$ or $\alpha(\beta\gamma)$ is defined, then both are, and they are equal.
b) If $\alpha\beta$ and $\beta\gamma$ are defined and β is an identity, then $\alpha\gamma$ is defined.
c) For each α there are identities ε_L and ε_R such that $\varepsilon_L\alpha$ and $\alpha\varepsilon_R$ are defined.
d) For each α there is an α^{-1} such that $\alpha^{-1}\alpha$ is a right identity for α and $\alpha\alpha^{-1}$ is a left identity for α.

A groupoid is called *connected* if it also satisfies:
e) Given any identities ε and ε', there is an element α such that $(\varepsilon\alpha)\varepsilon'$ is defined.

There are a number of immediate consequence of these axioms. First of all, $\alpha\beta$ is defined if and only if each right identity for α is a left identity for β. Consequently, if $\alpha\beta$ and $\beta\gamma$ are defined, then $\alpha(\beta\gamma)$ is defined. If ε and ε' are identities, then $\varepsilon\varepsilon'$ is defined if and only if $\varepsilon = \varepsilon'$, and then $\varepsilon^2 = \varepsilon$. Thus, the left and right identities for any element are unique. The cancellation laws follow easily, and therefore the uniqueness of the inverse. Another interesting remark is that any element satisfying $\alpha^2 = \alpha$ is an identity.

These remarks imply that a groupoid can be defined as a small category in which every morphism is an isomorphism. The objects of this category form a set X whose elements are the identities. If x and y are objects, hom (x, y) is the set of α having right identity x and left identity y.

1.6 Definition. A *topological groupoid* is a groupoid with a topology such that composition and taking of inverses are continuous.

1.7 Proposition. *Let $\mathscr{G}$ be a pseudogroup of homeomorphisms of X, let $\Gamma(\mathscr{G})$ be the set of germs of elements of $\mathscr{G}$, and let ι_x be the germ of the identity map at x. Let $\Gamma(\mathscr{G})$ have the topology generated by those sets of the form $\{f_x\}$ where f is a fixed element of $\mathscr{G}$ and x ranges over some open set contained in the domain of f. Let a composition in $\Gamma(\mathscr{G})$ be defined by $g_y f_x = (gf)_x$, which makes sense if and only if $f_x(x) = y$. Then*

$\Gamma(\mathcal{G})$ is a topological groupoid, and the map which takes x to ι_x is a homeomorphism of X onto an open subset of $\Gamma(\mathcal{G})$. If $f_x(x) = y$, then $(f_x)^{-1} = (f^{-1})_y$. $\Gamma(\mathcal{G})$ characterizes $\mathcal{G}$, in the sense that a homeomorphism between open subsets of X belongs to $\mathcal{G}$ if and only if its germ at each point belongs to $\Gamma(\mathcal{G})$. $\mathcal{G}$ is transitive if and only if $\Gamma(\mathcal{G})$ is connected.

Proof. Most of these statements are immediate consequences of the definitions. To prove that $\Gamma(\mathcal{G})$ characterizes $\mathcal{G}$, it is necessary to observe that if f and g belong to $\mathcal{G}$, then $\{x \mid f_x = g_x\}$ is open in X. This and the gluing axiom 1.1e) provide the proof. Note that $\{x \mid f(x) = g(x)\}$ is in general a larger set, which is closed, and that $\Gamma(\mathcal{G})$ is not in general a Hausdorff space. ▯

In principle, one could begin with a topological groupoid Γ, construct the pseudogroup of local homeomorphisms of its space of identities, and try to identify a subpseudogroup whose germs form a groupoid isomorphic to Γ. This procedure seems to be more difficult, and will not be discussed further, since it is not needed here.

The method for using $\mathcal{G}$ to construct spaces will now be described. Suppose V_μ are open sets in the model space for $\mu \in \mathcal{M}$, $\mathcal{M}$ some indexing set, $'U_\mu$ are sets disjoint from each other and from the model space, and $'f_\mu$: $'U_\mu \to V_\mu$ are homeomorphisms. Some equivalence relation is given on $\bigcup_{\mu \in \mathcal{M}} {}'U_\mu$ so that the quotient mapping q_μ: $'U_\mu \to X$ is one-one, where X is the quotient space. If $U_\mu = q_\mu('U_\mu)$, then there is a uniquely defined continuous mapping f_μ: $U_\mu \to V_\mu$ such that $f_\mu \circ q_\mu = {}'f_\mu$ is one-one. It is further hypothesized that $q_\mu^{-1}(U_\mu \cap U_\nu)$ is open in $'U_\mu$ for all $(\mu, \nu) \in \mathcal{M} \times \mathcal{M}$, so that f_μ is a homeomorphism. X need not be Hausdorff, even if the model space is. The final hypothesis is that

$$g_{\nu\mu} = f_\nu \circ f_\mu^{-1}: f_\mu(U_\mu \cap U_\nu) \to f_\nu(U_\mu \cap U_\nu) \tag{1.8}$$

are elements of $\mathcal{G}$ satisfying

$$g_{\nu\mu} \circ g_{\mu\lambda} = g_{\nu\lambda} \tag{1.9}$$

Equation (1.9) is known as the cocycle condition.

An important case of this construction occurs as follows: The given data are a topological space Y, an open covering $\{U_\mu \mid \mu \in \mathcal{M}\}$ of Y, a collection of open sets $\{V_\mu \mid \mu \in \mathcal{M}\}$ of the model space X, and a collection of homeomorphisms f_μ: $U_\mu \to V_\mu$ such that the $g_{\nu\mu}$ defined by Eq. (1.8) belong to $\mathcal{G}$ and satisfy (1.9). Then $'U_\mu$, $'f_\mu$, and q_μ could be found such that all the hypotheses are satisfied, but in practice there is no need to introduce them. The collection $\{f_\mu: U_\mu \to V_\mu \mid \mu \in \mathcal{M}\}$ is called a $\mathcal{G}$-atlas on Y.

1.10 Definition. If F: $Y \to Z$ is a continuous, open mapping, $\{f_\mu: U_\mu \to V_\mu \mid \mu \in \mathcal{M}\}$ is a $\mathcal{G}$-atlas on Y, and $\{g_\nu: W_\nu \to V_\nu \mid \nu \in \mathcal{N}\}$ is a $\mathcal{G}$-atlas on Z, then F is called a *$\mathcal{G}$-mapping* provided

$$g_\nu \circ F \circ f_\mu^{-1}: f_\mu(F^{-1}(W_\nu) \cap U_\mu) \to g_\nu(w_\nu \cap F(U_\mu))$$

belongs to $\mathcal{G}$. F is a *$\mathcal{G}$-isomorphism* provided F^{-1} exists and is also a $\mathcal{G}$-mapping.

1.11 Definition. A *$\mathscr{G}$-structure* on X is the class of all $\mathscr{G}$-atlases on X that are $\mathscr{G}$-isomorphic to a given $\mathscr{G}$-atlas. A *$\mathscr{G}$-space* is a space with a $\mathscr{G}$-structure.

Note that a $\mathscr{G}$-structure can be broken up into subclasses consisting of $\mathscr{G}$-equivalent atlases, where two atlases are said to be $\mathscr{G}$-equivalent if the identity mapping is a $\mathscr{G}$-isomorphism. The union of a $\mathscr{G}$-equivalence class of atlases is a maximal $\mathscr{G}$-atlas.

The definition of structure can of course also be given in the context of groupoids. Then the natural condition on $g_{\nu\mu}$ is that each of its germs can somehow be identified with an element of the groupoid. In the case of $\Gamma(\mathscr{G})$, one expects to get the same notion of structure as in the case of $\mathscr{G}$, as long as the mappings f_μ are required to be homeomorphisms. (A proof of this statement for the case of foliations is given in § 2.) By using continuous f_μ and an arbitrary topological groupoid, a more general kind of structure can be defined. Discussion of this will be deferred until Chapter III, where these structures occur naturally in the theory of classifying spaces for groupoids.

1.12 Example. The pseudogroup $\mathscr{G}$ defines a canonical $\mathscr{G}$-structure on its model space.

1.13 Definition. $\mathscr{H}$ is a *subpseudogroup* of $\mathscr{G}$ provided $\mathscr{H}$ is a pseudogroup with the same model space as $\mathscr{G}$, and every mapping which belongs to $\mathscr{H}$, also belongs to $\mathscr{G}$. A *subgroupoid* of Γ is a subset which is also a groupoid, and which contains the set of identities of Γ.

For the reasons discussed in the preface, examination of the notions of quotient pseudogroup and quotient groupoid will be deferred until the next section, where they can be discussed in the context of foliations.

A given topological space may or may not admit a $\mathscr{G}$-structure, which may or may not be unique. In Chapter III it will be shown how the question of existence of $\mathscr{G}$-structures on a given space Y can be reduced to problems in algebraic topology, at least for some $\mathscr{G}$ and some spaces. In this book Y will usually be a manifold, that is, a locally euclidean, paracompact, Hausdorff space. Any manifold can be written as the disjoint union of (arbitrarily many) connected open submanifolds. It is useful to note that for a connected, locally euclidean, Hausdorff space, the properties of paracompactness, σ-compactness, and second countability are equivalent. Most $\mathscr{G}$-structures on manifolds will be built using pseudogroups in euclidean space. A homeomorphism defined on an open subset of $\mathbb{R}^m$ into $\mathbb{R}^m$ always has an open image, so that there is no need to verify this condition when dealing with pseudogroups in $\mathbb{R}^m$. Furthermore, if $f \in \mathscr{G}$ and all the elements of the pseudogroup are differentiable, then the matrix $J^1(f)$ of first partial derivatives $\partial f^i/\partial x^j$ is nonsingular by condition c) of Definition 1.1. Identifying $\mathbb{R}^m$ with its tangent space implies that $J^1(f)$ is identified with an element of the general linear group $GL_m(\mathbb{R})$. For each point $x \in \mathbb{R}^m$

$$G_x = \{J^1(f) \mid f \in \mathscr{G}, f(x) = x\} \tag{1.14}$$

is a subgroup of $GL_m(\mathbb{R})$ called the linear isotropy group. If $\mathscr{G}$ is transitive, the subgroup G_x is conjugate to G_y by $J^1(h)$, where $h(x) = y$.

1.15 Definition. If $\mathscr{G}$ is transitive on $\mathbb{R}^m$ and 0 is the origin, the closure of the group G_0 defined by (1.14) is called the *Lie group associated to* $\mathscr{G}$, and denoted by $G(\mathscr{G})$.

The group $G(\mathscr{G})$ has a significance in bundle theory. Namely, if a manifold has a $\mathscr{G}$-structure, then the group of its tangent bundle is reducible to $G(\mathscr{G})$. As was mentioned in the preface, the converse is false, and there are many unsolved problems in this area. The nature of these problems will be discussed in § 3.

1.16 Definition. Let $\mathscr{G}$ be a pseudogroup defined on an open subset U in $\mathbb{R}^m$ and v be a vector field defined on U. v is called an *infinitesimal automorphism* of $\mathscr{G}$ if each diffeomorphism of an open subset of U into U defined by integrating v is an isomorphism between the restrictions to its domain and range of the canonical *$\mathscr{G}$-structure on U.*

The infinitesimal automorphisms play an important role in the theory of pseudogroups on manifolds, since they are the linearizations of automorphisms, hence yield partial differential equations that must be satisfied by local automorphisms.

The following examples include most of the important pseudogroups in $\mathbb{R}^m$. Most proofs are omitted, and some of the examples are discussed very briefly. More details can be found in the books of Sternberg [1964] and Kobayashi [1972]. Also, the methods developed later in this book are applicable to some of these examples.

1.17 Example. The pseudogroup $\mathscr{G}^k_m$ consists of all diffeomorphisms of class C^k defined on open subsets of $\mathbb{R}^m$ into $\mathbb{R}^m$. If $k \neq 0$, $G(\mathscr{G}^k_m)$ is the entire group $GL_m(\mathbb{R})$. A $\mathscr{G}^k_m$-space is a manifold of dimension m and differentiability class C^k, and a $\mathscr{G}^k_m$-structure is a differentiable structure on a manifold. Notice that even though $\mathbb{R}^1$ has only one $\mathscr{G}^k_1$-structure, there exist inequivalent smooth atlases. For example, the homeomorphism defined by $h(x) = x^3$ induces an isomorphism of inequivalent atlases. If k is infinite, every C^k vector field is an infinitesimal automorphism. If k is finite, there is no easy characterization of the vector fields which generate C^k diffeomorphisms. A detailed study of the problems involved is contained in § 4.

1.18 Example. Let $m = 2n$ and consider $\mathbb{R}^m$ as an underlying space of the space $\mathbb{C}^n$ of n complex variables. Let $\mathscr{G}^{\mathbb{C}}_n$ consist of all complex analytic homeomorphisms defined on open subsets of $\mathbb{C}^n$ into $\mathbb{C}^n$. These form a pseudogroup because $J^1(f)$ for any such f is necessarily nonsingular. $G(\mathscr{G}^{\mathbb{C}}_n)$ is $GL_n(\mathbb{C})$, a $\mathscr{G}^{\mathbb{C}}_n$-space is a complex analytic manifold, and a $\mathscr{G}^{\mathbb{C}}_n$-structure is a complex analytic structure on a manifold. Here $GL_n(\mathbb{C})$ is considered as a subgroup of $GL_{2m}(\mathbb{R})$, namely, the image of the monomorphism

$$C \to \begin{pmatrix} C_1 & -C_2 \\ C_2 & C_1 \end{pmatrix}$$

where C_1 and C_2 are real $n \times n$ matrices such that $C = C_1 + iC_2$. The image may be described as the set of matrices A such that $A^{-1}JA = J$ where

$$J = \begin{pmatrix} 0 & -I \\ I & 0 \end{pmatrix}$$

and I is the $n \times n$ identity matrix. The fact that $J^1(f)$ belongs to this subgroup says precisely that f satisfies the n-dimensional Cauchy-Riemann equations, hence is both complex and real analytic. The infinitesimal automorphisms of $\mathscr{G}_n^{\mathbb{C}}$ are vector fields of the form

$$v^i(z^1, \ldots, z^n)\, \partial/\partial z^n$$

where $(z^1, \ldots, z^n)$ are coordinates in $\mathbb{C}^n$ and the v^i are complex analytic functions of their arguments. $\mathscr{G}_n^{\mathbb{C}}$ is contained in $\mathscr{G}_{2n}^{\omega}$, and $\mathscr{G}_{2n}^k$ defines a morphism from $\mathscr{G}_n^{\mathbb{C}}$ to $\mathscr{G}_{2n}^k$.

The pseudogroup $\mathscr{G}_n^{\mathbb{C}}$ is transitive. Some other important transitive pseudogroups are the volume preserving pseudogroup, the contact pseudogroup, and the symplectic pseudogroup. The latter two are extremely important in mechanics, in which context symplectic transformations are also known as Hamiltonian or canonical transformations. For a transitive pseudogroup $\mathscr{G}$, any two $\mathscr{G}$-spaces are locally isomorphic. Because an isometry preserves curvature, it is not true that any two Riemannian manifolds are locally isomorphic. Hene, to deal with Riemannian geometry in this context, it is necessary to use a pseudogroup which is not transitive. Moreover, the model space must be very large to include all the possible local behaviours. It is in fact the sheaf of germs of Riemannian metrics on $\mathbb{R}^m$, which will now be defined.

1.19 Example. A C^k Riemannian metric g_U on an open set U in $\mathbb{R}^m$ is a nondegenerate quadratic form on the tangent space at each point of U, such that the coefficients in terms of the standard dual basis vary in a C^k manner from point to point. The formula is

$$g_U(x^1, \ldots, x^m) = \sum_{i,j=1}^{m} g_{ij}(x^1, \ldots, x^m)\, dx^i\, dx^j$$

where $g_{ji} = g_{ij}$ is a smooth function on U. Two such are equivalent at $x \in U \cap V$ if there is a neighborhood W of x contained in $U \cap V$ such that $g_U | W = g_V | W$. An equivalence class at x is called a germ of Riemannian metrics at x, and is denoted by g_x. The set of all germs is topologized by taking the smallest topology containing all the sets of the form $T\{g_U\} = \{g_x \,|\, \text{for some } x \in U, g_x \text{ is the germ of } g_U \text{ at } x\}$. There is a natural projection π of the sheaf onto $\mathbb{R}^m$ which takes g_x into x, and which induces a homeomorphism $\pi\{g_U\}$ between $T\{g_U\}$ and U. This makes the sheaf into a manifold for which the transition function $\pi\{g_V\} \circ \pi\{g_U\}^{-1}$ is the identity map of $U \cap V$. Hence, the sheaf is a C^{ω}-manifold, whatever k may be. It is not a Hausdorff space, however, unless $k = \omega$. The pseudo-group $\mathscr{R}_m$ consists of diffeomorphisms f of class C^{k+1} defined an open sets of the sheaf, which are isometries in the sense that the maps

$$\pi\{g_V\} \circ f \circ \pi\{g_U\}^{-1}$$

are isometries wherever they are defined. To every germ is associated a subgroup of the orthogonal group 0_m, namely, the group of Jacobians of isometries of representatives g_U of the given germ. The groups associated with different germs are not in general conjugate. An $\mathscr{R}_m$-space is called a Riemannian manifold, and an $\mathscr{R}_m$-structure is called a Riemannian structure. The infinitesimal automorphisms are obtained by combining infinitesimal rotations and infinitesimal translations. They are called infinitesimal isometries or Killing vector fields.

1.20 Example. Let G be any group of transformations of $\mathbb{R}^m$. This means that G is a topological group and there is a C^0-mapping

$$\phi: G \times \mathbb{R}^m \to \mathbb{R}^m$$

such that $\phi(g, \ )$ is a C^k diffeomorphism and $\phi(g_1, \phi(g_2, x)) = \phi(g_1 g_2, x)$. If $k \geqq 1$, then G must be a Lie group. The collection of all restrictions to open subsets of all maps $\phi(g, \ )$ is a pseudogroup $\mathscr{L}(G)$ called the localization of G. A case of particular interest is the affine group $A_m(\mathbb{R})$, for which there is an exact sequence

$$1 \to \mathbb{R}^m \to A_m(\mathbb{R}) \rightleftarrows GL_m(\mathbb{R}) \to 1.$$

$A_m(\mathbb{R})$ is the group of automorphisms of the affine structure of $\mathbb{R}^m$, that is, mappings which for every k, take every family of cosets of a k-dimensional vector subspace to a family of cosets of a k-dimensional vector subspace. In particular, $\mathbb{R}^m$ acts on itself by translations and so may be considered as a subgroup of $A_m(\mathbb{R})$; the formula is simply $\phi(a, x) = a + x$. The exact sequence splits, since $GL_m(\mathbb{R})$ may be considered as the subgroup which leaves the origin fixed. The corresponding pseudogroup will be denoted by $\mathscr{L}(A)$, and $G(\mathscr{L}(A)) = GL_m(\mathbb{R})$. The same construction may be applied to any subgroup of $A(m; \mathbb{R})$, but the example of the affine transformations which preserve the standard contact structure of $\mathbb{R}^m$ shows that a subgroup can be transitive without containing all the translations. For any subgroup F of $A(m; \mathbb{R})$, the corresponding $G(\mathscr{L}(F))$ is found by taking the quotient of F by its intersection with $\mathbb{R}^m$. This method of finding the associated linear group works only for subgroups of the affine group, since the affine group is the largest subgroup of the homeomorphism group of $\mathbb{R}^m$ of which the translation group is a normal subgroup. The assumption of existence of an $\mathscr{L}(G)$-structure is a rather strong hypothesis. For example, if G is the group of rigid motions of euclidean space with its usual metric then an $\mathscr{L}(G)$-structure is a flat Riemannian structure.

1.21 Example. If G is a subgroup of $GL_m(\mathbb{R})$ and $k \geqq 1$, $\mathscr{G}_m^k(G)$ is defined to be the subset of $\mathscr{G}_m^k$ consisting of mappings whose Jacobian matrix at each point belongs to G. The exact sequence of the preceding example associates to G a subgroup $A(G)$ in $A(m; \mathbb{R})$, and the localization $\mathscr{L}(A(G))$ is contained in $\mathscr{G}_m^\omega(G)$. The pseudogroup $\mathscr{G}_n^{\mathbb{C}}$ is the same as $\mathscr{G}_{2n}^\omega(GL_n(\mathbb{C}))$, while $\mathscr{G}_m^k$ is the same as $\mathscr{G}_m^k(GL_m(\mathbb{R}))$. Similar remarks hold for the volume preserving and symplectic pseudogroups, but the construction fails for the contact pseudogroup since the projections into $GL_m(\mathbb{R})$ of the isotropy groups at different points are conjugate rather than identical. In most cases, the elements of $\mathscr{G}_m^k(G)$ also satisfy certain conditions on their higher order derivatives. These conditions will be studied in Chapter II.

The construction of $\mathscr{G}_m^k(G)$ may also be phrased in terms of bundle theory. An isomorphism between the tangent frame bundle of $\mathbb{R}^m$ and $\mathbb{R}^m \times GL(m, \mathbb{R}^m)$ is defined by taking as a basis in $\mathbb{R}^m$ the coordinate vectors having one entry 1 an all other entries 0. Then $\mathbb{R}^m \times \mathrm{G}$ can be considered as a reduction of the group of the frame bundle to G, and as such is called the standard G-structure on $\mathbb{R}^m$. $\mathscr{G}_m^k(G)$ consists precisely of the local automorphisms of the standard G-structure on $\mathbb{R}^m$.

Historically, the notation of pseudogroup arose during a long process of trying to understand the foundations of geometry. The notion of group was introduced in an algebraic context by Galois [1846], and by the turn of the century (Poincaré [1921], written before 1912) mathematicians were considering a dual dichotony of finite versus infinite and discrete versus continuous groups. The work of Galois involved finite discrete groups, while infinite discrete groups were found in the study of Fuchsian groups and related questions. The continuous groups were not groups in the modern sense, but sets of transformations with variable domains. In this context, finite meant finite dimensional. Thus, a finite continuous group was what is now called a neighborhood of the identity in a Lie group. The modern global point of view was developed slowly in the work of Lie, Engel, Killing, and Elie Cartan. The implications for geometry were formulated by Klein [1872] in his famour Erlangen inaugural dissertation. For a long time, virtually everything known about infinite (dimensional) continuous groups was due to Cartan [1904, 1908, 1909], who followed Lie in using the partial differential equations for the elements as part of the definition for infinite group. His geometric theory of partial differential equations [1901] was used to study the local properties of these groups. The name pseudogroup was introduced by Veblen and Whitehead [1932], but they leave the partial differential equations out of the definition, since their purpose is to construct manifolds. Shortly afterwards, Whitney [1935, 1937] introduced fiber spaces in a successful effort to capture some of the global properties of manifolds.

The idea of groupoid also arose first in a purely algebraic context, in the study of the composition of quadratic forms by Brandt [1913, 1927]. Its usefulness in the present context was recognized by Nijenhuis [1952], who identified local differential geometry with the invariant theory of the groupoid $\Gamma(\mathscr{G}_m^\infty)$, but did not explicitly mention its topology. The latter was used by Haefliger [1970] in constructing a classifying space for $\Gamma(\mathscr{G}_m^\infty)$.

Recently there has also been some success obtained by treating the group of diffeomorphisms of a manifold as a topological group, whose Lie algebra (in some sense) is the space of vector fields on the manifold. This work is mostly beyond the scope of the present book, and will be ignored or referred to in passing.

There is much that remains to be understood about structures on manifolds, but there is some indication that events will justify the speculation of J. H. C. Whitehead [1952] that the work on infinite continuous (pseudo) groups may be Elie Cartan's best.

2. Foliations

Now that the general idea of structure on a manifold has been introduced, and explained by means of a number of examples, it is time to consider the example of foliations, which is the main object of study in this book. This type of structure will be defined and related to the natural notions of quotient and substructure on a manifold, that is, of epimorphism and monomorphism in the language of category theory. A number of classes of examples of foliated manifolds will be given, after which some important objects associated to a foliation will be defined. Finally, the question of refining the structure by imposing additional tangent or transverse conditions will be discussed.

Suppose that $\mathbb{R}^m = \mathbb{R}^p \times \mathbb{R}^q$ has coordinates $(x^1, \ldots, x^p, y^1, \ldots, y^q)$ and consider $\mathbb{R}^m$ as the union of affine subspaces defined by $y^1 = c^1, \ldots, y^q = c^q$ for constants $c^1, \ldots, c^q$. Each such subspace is called a leaf.

2.1 Definition. The pseudogroup $\mathscr{G}^k_{p;\,m}$ consists of all elements of $\mathscr{G}^k_m$ such that the intersection of any convex subset of the domain with a leaf is mapped into a leaf.

It follows that if $g \in \mathscr{G}^k_{p;\,m}$ has a convex domain, then it has a formula of the form

$$g(x, y) = (g_1(x, y), g_2(y)) \tag{2.2}$$

where g_1 maps into $\mathbb{R}^p$ and g_2 maps into $\mathbb{R}^q$. If $k \geqq 1$, then an equivalent condition is the requirement that the tangent map induced by g takes the tangent space of a leaf into the tangent space of a leaf. The group $G(\mathscr{G}^k_{p;\,m})$ consists of all linear transformations which map the subspace defined by $y = 0$ into itself. The matrix with respect to the coordinate basis of any such transformation has the form

$$\begin{pmatrix} A & B \\ 0 & D \end{pmatrix} \tag{2.3}$$

where A is $p \times p$ nonsingular and D is $q \times q$ nonsingular. A is a matrix for the induced linear transformation on the subspace $y = 0$, and D is a matrix for the quotient transformation.

2.4 Definition. A $\mathscr{G}^k_{p;\,m}$-structure is called a *foliation of dimension p, codimension q, and class C^k*. A manifold with a $\mathscr{G}^k_{p;\,m}$-structure is called a *foliated manifold*. The group consisting of matrices of the form (2.3) is denoted by $G_{p;\,m}(\mathbb{R})$.

The definition of $\mathscr{G}$-structure (Definition 1.11) requires the consideration of atlases for a $\mathscr{G}^k_{p;\,m}$-structure such that a single equation of the form (2.2) holds throughout each overlap of a pair of coordinate neighborhoods. The following Proposition states that it is sufficient to require (2.2) on the neighborhood of each point, and then an atlas satisfying the stronger condition and having a number of other useful properties can be found.

2.5 Proposition. *Let h_μ: $U_\mu \to P_\mu \times Q_\mu$, $\mu \in \mathcal{M}$, be a family of coordinate charts covering the foliated manifold M such that each P_μ is connected. Suppose further that P_μ and Q_μ are open sets in $\mathbb{R}^p$ and $\mathbb{R}^q$ respectively, and that corresponding to each point of any intersection of two U_μ, there is a neighborhood in which a coordinate change of the form* (2.2) *is valid. Then there is a family of charts h_ν: $V_\nu \to P_\nu \times Q_\nu$, $\nu \in \mathcal{N}$, $\mathcal{M} \cap \mathcal{N} = \emptyset$, and a map θ: $\mathcal{N} \to \mathcal{M}$ with the following properties:*

(i) *P_ν is connected.*
(ii) *$\bar{P}_\nu$ is a compact subspace of $P_{\theta\nu}$ and $\bar{Q}_\nu$ is a compact subspace of $Q_{\theta\nu}$.*
(iii) *h_ν is the restriction of $h_{\theta\nu}$.*
(iv) *If $V_\nu \cap V_\lambda \neq \emptyset$, then $\bar{V}_\nu \subset U_{\theta\lambda}$.*
(v) *$\{V_\nu\}$ is a locally finite family and each $\bar{V}_\nu$ is compact.*
(vi) *For each pair $(\nu, \lambda) \in \mathcal{N} \times \mathcal{N}$, let $A_{\nu\lambda}$ be the subset of Q_ν obtained by projecting $h_\nu(V_\nu \cap V_\lambda)$ along P_ν. Then there exists a unique onto homeomorphism $g_{\nu\lambda}$: $A_{\lambda\nu} \to A_{\nu\lambda}$ which is the projection of $h_\nu h_\lambda^{-1}$.*

Proof. This is proved by Epstein, Millet, and Tischler [1977]. □

Note that in the transformation rule (2.2) the components g_2 are themselves elements of $\mathcal{G}_q^k$. Thus, associated with the foliation is some kind of quotient structure. In fact, this structure is enough to determine the foliation, in a sense which will now be made precise. Suppose $\{f_\mu$: $U_\mu \to V_\mu\}$ is an atlas for the foliation, and π: $\mathbb{R}^m \to \mathbb{R}^q$ is defined by $\pi(x, y) = y$. If $k \geqq 1$, each map

$$\pi_\mu = \pi \circ f_\mu\colon U_\mu \to W_\mu = \pi(V_\mu)$$

is a submersion, that is, $J^1(\pi_\mu)$ has rank q. If $P \in U_\mu \cap U_\nu$, then for any convex neighborhood W_P of $\pi_\mu(P)$ contained in $\pi_\mu(U_\mu \cap U_\nu)$ there is a homeomorphism of class C^k of W_P into W_ν, induced by $g_{\nu\mu}$ and denoted by $g_{\nu\mu P}$. These mappings satisfy the cocycle condition in the following form:

If $g_{\mu\lambda}(\pi_\lambda(P)) = \pi_\mu(Q)$, then $g_{\nu\mu Q} \circ g_{\nu\mu P} = g_{\nu\lambda P}$ at all points where the left hand side is defined. The cocyle conditions in particular means that $g_{\mu\mu P}$ is the identity on its domain.

Conversely, suppose given a covering $\{U_\mu\}$ of a manifold M, open sets W_μ in $\mathbb{R}^q$, submersions π_μ: $U_\mu \to W_\mu$, and homeomorphisms $g_{\nu\mu P}$ of class C^k, defined for $P \in U_\mu \cap U_\nu$, such that the domain of $g_{\nu\mu P}$ is an open neighborhood W_P of $\pi_\mu(P)$ in $\pi_\mu(U_\mu \cap U_\nu)$, the range lies in W_ν, and the cocycle condition is satisfied. Then by elementary calculus, for any P there are a neighborhood U_P of P such that $\pi_\mu(U_P) \subset W_P$ and a C^k-homeomorphism f_P: $U_P \to \mathbb{R}^{p+q}$ such $\pi \circ f_P = \pi_\mu$. Then $\{f_P$: $U_P \to f_P(U_P)\}$ is $\mathcal{G}_{p;\,m}^k$-atlas, by the cocycle condition.

In case $k = 0$, the conclusion of this theorem from calculus is taken as the definition of submersion. Hence, in this case also the two approaches to defining foliations are equivalent. In any case, the submersions π_μ do not form a $\mathcal{G}_q^k$-structure if $p > 0$, since they are not homeomorphisms. This is an example of a $\Gamma(\mathcal{G}_q^k)$-structure, which will be introduced in Chapter III.

In each U_μ, the sets defined by $y^1 = c^1, \ldots, y^q = c^q$ are embedded submanifolds. In order to study the geometry of foliations, it is important to see how these sets join together in passing from U_μ to U_ν. In order to do this, a finer topology called the leaf topology is introduced, namely, the topology generated by all sets of the form $U \cap \pi_\mu^{-1}(x)$ where U is open U_μ and $x \in W_\mu$. In this topology, the foliated manifold is p-dimensional and paracompact, but no longer admits a countable base for the open sets.

2.6 Definition. A *leaf* of a foliation is a component of the manifold in the leaf topology.

Each leaf is a p-dimensional, σ-compact, connected manifold with a countable base. The injection mapping of a leaf into the original topology is continuous and one-one, and if $k \geqq 1$, the mapping has rank p. The image is not necessarily a closed set in the original topology. An isomorphism of foliations induces a homeomorphism of the leaf topologies, hence takes each leaf onto a leaf. On the other hand, the leaves of a given foliation may be of quite distinct topological types.

The importance of foliations in the general theory of local structures on manifolds arises from the fact that they can be considered as the local quotient structures. This remark has already been partially justified in the Preface, and by the detailed discussion of local submersions just given. The insufficiency of the notion of global quotient is illustrated by the foliation of the 2-torus by a family of dense spirals: the quotient space has uncountably many points, but only two open sets. The solution of many specific problems in foliation theory depends upon finding a substitute for the quotient space appropriate to the problem. Some examples are the normal bundle and various kinds of connections and metrics. The theory of Γ-structures, which will be discussed in Chapter III, casts further light on the notion of foliation as a quotient structure.

In order to understand the definitions better, it is appropriate to look at some simple examples. Further examples will be given in § 5 and § 6, after more technique has been developed.

2.7 Example. Let M_1 and M_2 be manifolds, and $\pi\colon M_1 \times M_2 \to M_2$ be the projection. Then π defines a foliation, such that each leaf is a component of $M_1 \times \{y\}$ for some $y \in M_2$.

2.8 Example. Let $\pi\colon M \to N$ be a fiber bundle. Then π defines a foliation, such that each leaf is a component of some fiber.

2.9 Example. Let M be a manifold with a nonzero vector field. Then the existence theory for ordinary differential equations implies that M has a 1-dimensional foliation. On the two dimensional torus, there are vector fields such that every leaf is dense in the whole manifold. In this example the leaf topology induces on each leaf a different topology from that induced by the topology of the original manifold. There are also vector fields such that some leaves are circles while some other leaves are immersed lines spiraling in to some set of circles. In other cases, there may be leaves that spiral in to very complicated closed sets [Denjoy, 1932]. Many other examples of complicated global behaviour may be found in standard works on ordinary differential equations.

2.10 Example. Let $\pi\colon E \to M$ be a vector bundle with q-dimensional fiber. Since $GL_q(\mathbb{R})$ is contained in the group of diffeomorphisms of $\mathbb{R}^q$ that leave the origin fixed, the group of any vector bundle may be extended to the latter group. Suppose that after this extension is made, the bundle is equivalent to a bundle with the extended group, but with the discrete topology on that group. Then there is an open covering $\{U_\mu\}$ of M and homeomorphisms $F_\mu\colon \pi^{-1}(U_\mu) \to U_\mu \times \mathbb{R}^q$ such that the composition $F_\nu \circ F_\mu^{-1}\colon (U_\mu \cap U_\nu) \times \mathbb{R}^q \to (U_\mu \cap U_\nu) \times \mathbb{R}^q$ is constant as a function of the first factor on each component of $U_\mu \cap U_\nu$, and is a differmorphism on the second factor leaving the origin fixed. These product structures give rise to submersions $\pi_\mu\colon \pi^{-1}(U_\mu) \to \mathbb{R}^q$, while the $F_\nu \circ F_\mu^{-1}$ are elements of $\mathscr{G}_q^k$. Hence, E has a foliation transverse to the fibers, such that the zero section is a leaf. In general, the existence of foliation transverse to the fibers does not imply the discreteness of the extended structure group. The meaning of such foliations will be discussed further in connection with Example 5.3.

Since $G(\mathscr{G}^k_{p;\,m}) = G_{p;\,m}(\mathbb{R})$, the group of the tangent bundle of a foliated manifold is reducible to $G_{p;\,m}(\mathbb{R})$. This fact can be made explicit in a very geometric way. The tangent space T_P to the manifold at the point P has a distinguished subspace $T_P(L)$, the tangent space to the leaf through P. Consider all those bases for T_P such that the first p vectors belong to $T_P(L)$. As P varies, these form a subset of the frame bundle which is itself a principal $G_{p;\,m}(\mathbb{R})$ bundle. The collection of spaces $T_P(L)$ forms a p-dimensional vector bundle which is a subbundle of the tangent bundle of the manifold.

2.11 Definition. The collection of spaces $T_P(L)$ with its natural vector bundle structure is called the *tangent bundle of the foliation* $\mathscr{F}$, and is denoted by $\tau(\mathscr{F})$. Its quotient bundle is called the *normal bundle of the foliation* and denoted by $\nu(\mathscr{F})$.

There is a one-one correspondence between subbundles of dimension p and reductions to $G_{p;\,m}(\mathbb{R})$, but the existence of either does not imply the existence of a foliation.

Since every sequence of vector bundles splits, the normal bundle is isomorphic to a subbundle of the tangent bundle $\tau(M)$, such that $\tau(M)$ is the direct sum of the tangent and normal bundles of $\mathscr{F}$. The choice of complement is not canonical, unless some additional structure is given. Such a choice is equivalent to a further reduction of the structural group to the direct sum

$$GL_p(\mathbb{R}) \oplus GL_q(\mathbb{R})$$

It may not be possible to choose the complement so that it is tangent to a foliation, that is, the symmetry at the tangent bundle level does not necessarily carry over to the pseudogroup level.

2.12 Definition. Let N be a foliated manifold, with the foliation $\mathscr{F}$ defined by local submersions $\pi_\mu\colon U_\mu \to \mathbb{R}^q$. A mapping $f\colon M \to N$ is said to be *transverse to* $\mathscr{F}$ provided each of the compositions $\pi_\mu \circ f$ is a submersion of $f^{-1}(U_\mu)$ into $\mathbb{R}^q$.

Then the local submersions $\pi_\mu \circ f\colon f^{-1}(U_\mu) \to \mathbb{R}^q$ define a foliation $f^*(\mathscr{F})$ on M.

2.13 Definition. Let $i\colon L \to M$ be the injection map of the leaf, taken with the leaf topology, and let $\nu(L) = i^* \nu(\mathscr{F})$. $\nu(L)$ is called the *normal bundle of the leaf.*

There exists an immersion f of the total space E of $\nu(L)$ into M such that each zero vector maps into its base point and each fiber is mapped transversally to $\mathscr{F}$. Hence $f^* \mathscr{F}$ is a foliation transverse to the fibers, such that the zero section is a leaf. If $k \geqq 2$, such a mapping f may be constructed by choosing a Riemannian metric on M. Then the exponential mapping along geodesics orthogonal to L embeds a neighborhood of the zero section with the required transversality property. f is obtained by composing with a fiber-preserving diffeomorphism of E into the neighborhood embedded by the exponential map. Clearly, many choices are involved. The intrinsic meaning of this construction and its relation to Example 2.10 will be discussed in connection with Example 5.3.

It is sometimes useful to consider subpseudogroups of $\mathscr{G}^k_{p;\,m}$. The most important ones arise from the fact that any element of $\mathscr{G}^k_{p;\,m}$ with convex domain has associated to it an element of $\mathscr{G}^k_q$, which can be called its $\mathscr{G}^k_q$-component. By requiring this element to belong to a given subpseudogroup $\mathscr{H}$ of $\mathscr{G}^k_q$, one obtains a subpseudogroup of $\mathscr{G}^k_{p;\,m}$.

2.14 Definition. A $\mathscr{G}^k_{p;\,m}$-structure is said to have a *transverse $\mathscr{H}$-structure* if the $\mathscr{G}^k_q$ component of each coordinate transformation belongs to a fixed subpseudogroup $\mathscr{H}$ of $\mathscr{G}^k_q$. It is said to have a *tangent $\mathscr{K}$-structure* if the restriction of each coordinate transformation to each leaf belongs to a fixed subpseudogroup $\mathscr{K}$ of $\mathscr{G}^k_p$.

In the first case, the group of $\nu(\mathscr{F})$ is reducible to $G(\mathscr{H})$, while in the second case, the group of $\tau(\mathscr{F})$ is reducible to $G(\mathscr{K})$.

2.15 Example. Let $\pi\colon M \to N$ be a fiber bundle, considered as a foliation as described in Example 2.8. If N is a complex manifold of complex dimension q, then this foliation has a transverse $\mathscr{G}^{\mathbb{C}}_q$-structure. If the fibers are complex manifolds of complex dimension p and the complex structures vary smoothly, then it has a tangent $\mathscr{G}^{\mathbb{C}}_p$-structure.

The roots of foliation theory can be found in many theories: ordinary differential equations, partial differential equations, level sets of functions, and others. Each of these examples is related to some aspect of the theory, but a global formulation sufficiently general to include all these cases is a recent creation. The leaf topology was introduced by Ehresmann and Reeb [1944], while the words "feuille" and "variété feuilletée" were introduced by Reeb [1947]. The early results of the theory are contained in the theses of Reeb [1952] and Haefliger [1956, 1958], written under Ehresmann's direction. Reeb proved the stability of compact, simply connected leaves and Haefliger the nonexistence of real analytic foliations of spheres. Novikov [1965] found some interesting properties of foliations on

3-dimensional manifolds. The effective use by Anosov [1962, 1963, 1967] of foliations in the study of dynamical systems created interest among a larger group of mathematicians, as did two developments of the late sixties: the existence theorems for foliations an open manifolds due to Phillips [1968, 1969, 1970], Gromov [1968, 1969], and Haefliger [1970, 1971], and the theorem of Bott [1970] on the vanishing of certain characteristic classes of the normal bundle of a foliation. Early work on the differential geometry of foliations was done by Walker [1955, 1958], Willmore [1956a, 1956b, 1957], Reinhart [1958, 1959a, 1959b, 1961], Hermann [1960, 1962], Vidal [1964], and Sacksteder [1965]. Furthermore, differential geometric methods found some use in the work of Anosov and Bott mentioned above. There will be no effort to summarize later developments here, since the rest of the book may be viewed as such a summary.

3. The Integrability Problem

Given a pseudogroup $\mathscr{G}$ and a manifold M, a fundamental problem is to determine whether M admits a $\mathscr{G}$-structure. If it does, a further problem is to classify the $\mathscr{G}$-structures up to some reasonable equivalence relation. These problems have proved to be extremely difficult and it is only for a few $\mathscr{G}$ that a complete solution exists. As has been remarked in the Preface, a necessary condition for the existence of a differentiable $\mathscr{G}$-structure is the existence of a reduction of the group of the tangent bundle to $G(\mathscr{G})$, but this is rarely sufficient. The case of $\mathscr{L}(\mathbb{R}^m)$, as defined in Example 1.20, is instructive, in that it illustrates the principal difficulties in a fairly direct way.

$\mathbb{R}^m$, as a Lie group , acts on itself by translations. $\mathscr{L}(\mathbb{R}^m)$ consists of all restrictions of these translations to open subsets of $\mathbb{R}^m$. $J^1(t)$ for any translation t is the identity matrix so $G\mathscr{L}(\mathbb{R}^m))$ is the group with one element. A reduction of the group of the tangent bundle to the identity is equivalent to finding m vector fields $\{v_1, \ldots, v_m\}$ independent at every point of M. These fields are called a framing, and M is called parallelizable if it admits a framing. There are many examples of parallelizable manifolds: every oriented 3-manifold [Stiefel 1936, Whitney 1941] and every Lie group, to name some classes of examples. On the other hand, the possession of an $\mathscr{L}(\mathbb{R}^m)$ structure is a much more restrictive condition. Indeed, a compact connected manifold which has an $\mathscr{L}(\mathbb{R}^m)$ structure is in a natural way an abelian Lie group.

3.1 Proposition. *Let M be a compact, connected manifold which admits an $\mathscr{L}(\mathbb{R}^m)$ structure. Then an analytic multiplication can be defined on M so that it becomes a quotient of $\mathbb{R}^m$ by a subgroup H which is discrete in its induced topology. There is an analytic diffeomorphism between M and the product of m copies of S^1, and the subgroup H is free abelian on m generators.*

Proof. Suppose the $\mathscr{L}(\mathbb{R}^m)$ structure is given by coordinate mappings $f_u\colon U_u \to V_u$ such that each $g_{\nu\mu}$ is a euclidean translation, where

$$g_{\nu\mu}\colon f_\mu(U_\mu \cap U_\nu) \to f_\nu(U_\mu \cap V_\nu).$$

Since translations are analytic diffeomorphisms, these data make M an analytic manifold. Given real constants a_j, the differential form $\Sigma a_j dx^j$ on $\mathbb{R}^m$ induces a form $\Sigma_j a_j dx^j_\mu$ in the coordinates of each V_μ. Since $g_{\nu\mu}$ is a translation and therefore leaves $\Sigma_j a_j dx^j$ invariant, these local formulas define differential forms $\Sigma a_j \theta^j$ on M, where θ^j is the form corresponding to dx^j. In a similar way, the vector field $\Sigma_i a^i \partial/\partial x^i$ on $\mathbb{R}^m$ gives rise to a vector field $\Sigma_i a^i X_i$ on M. It is clear from the local formulas that each such form is closed, that the bracket of any two such vector fields is 0, and that if either a form or a vector field vanishes at one point, then it is identically zero. Since M is compact, each vector field is complete, that is, its integral curves are defined for all values of the natural parameter. Moreover, these integral curves are the geodesies of the metric defined by $\Sigma(\theta^i)^2$.

3.2 Lemma. *Given any point y in M, there is a coordinate system f_y taking a neighborhood U_y of y into $\mathbb{R}^m$ such that the integral curves of $\Sigma a^i X_i$ are mapped into integral curves of $\Sigma a^i \partial/\partial x^i$. f_y is compatible with the original coordinate covering in the sense of $\mathscr{L}(\mathbb{R}^m)$ structures, so it is an analytic diffeomorphism. It is uniquely determined by the choice of $f_y(y)$ and the domain U_y.*

Proof. There is a μ such that $y \in U_\mu$. Given $f_y(y)$, define

$$f_y(z) = f_\mu(z) - f_\mu(y) + f_y(y)$$

for all z in some neighborhood of y contained in U_μ. This proves the existence of a coordinate system with the required properties. Since any two such coordinate systems differ by a translation, if they agree at a point and have the same domain, they are identical. □

The preceding lemma gives the local structure of the integral curves of the vector fields $\Sigma a^i X_i$, while the next lemma gives some global properties.

3.3 Lemma. *Let $X = \Sigma a^i X_i$, $Y = \Sigma b^i X_i$, and let P be a point of M. Then the following three paths end at the same point of M:*

(i) *The path starting at P, following the integral curve of X for a parameter change of 1, and then proceeding from the end point of the first segment along the integral curve of Y for a parameter change of 1.*
(ii) *The path starting at P and following first Y and then X in the analogous manner.*
(iii) *The path starting at P and following the integral curve of $X + Y$ through a parameter change of 1.*

Proof. In case $M = \mathbb{R}^m$ and $X_i = \partial/\partial x^i$, the lemma is wellknown. By the compactness of M and Lemma 3.2, there is an $\varepsilon \geq 0$ such that if $|X| < \varepsilon$, $|Y| < \varepsilon$, and $|X + y| < \varepsilon$, the lemma is true. For arbitrary X and Y, it suffices to choose n large enough so that $\left|\frac{1}{n} X\right| < \varepsilon$, $\left|\frac{1}{n} Y\right| < \varepsilon$ and $\left|\frac{1}{n}(X + Y)\right| < \varepsilon$, then apply the preceding remark to each small parallelogram. □

It is now possible to define the map $p\colon \mathbb{R}^m \to M$ which will have the desired properties. Choose a base point y_0 in M, and set $p(0) = y_0$. Map $(x^1, \ldots, x^m)$ into the point with parameter value 1 of the integral curve of $\Sigma x^i X_i$ beginning at y_0. If $y = p(x)$, consider a mapping f_y as described in Lemma 3.2, subject to the condition that $f_y(y) = x$. Then by Lemmas 3.2 and 3.3 $p(f_y(z)) = z$ for all z in some neighborhood of y contained in the domain of f_y. Hence, p is an analytic submersion, and in fact an analytic local diffeomorphism, with local inverse f_y. p is onto because any point of M can be joined to $p(0)$ by a geodesic.

3.4 Lemma. *$p(x) = p(y)$ if and only if $p(x-y) = p(0)$.*

Proof. If $p(x) = p(y)$, then p takes the segment joining x to y into a C^ω embedded circle. By commutativity of the vector fields, the flow which takes $p(y)$ to $p(0)$ takes this circle to a circle, so $p(x-y) = p(0)$. The converse is proved by the same method, using the flow from $p(0)$ to $p(y)$. □

It follows immediately from this lemma that $H = p^{-1}(p(0))$ is a subgroup whose cosets are the sets $p^{-1}(y)$. Since $\mathbb{R}^m$ is abelian, this implies that p induces the structure of a commutative group on M. Because p is a local diffeomorphism, H meets the ball of radius ε around 0 only at 0, provided ε is small enough. Hence, the balls of radius $\frac{1}{2}\varepsilon$ around any two points of H are disjoint, so that H is discrete in the induced topology. Furthermore, any line through 0 which meets H, does so in the set of integer multiples of a closest point to 0. Since H is a subgroup of $\mathbb{R}^m$, it is abelian and has no elements of finite order. Furthermore, the fundamental group of $\mathbb{R}^m/H$ is isomorphic to H, since $\mathbb{R}^m$ is contractible. The last sentence of Proposition 3.1 can now be proved by induction on m. Choose a subgroup $\mathbb{R}$ of $\mathbb{R}^m$ with $\mathbb{R} \cap H \neq \{0\}$, say $\mathbb{R} \cap H = H_1$ where H_1 is free cyclic. Then the exact sequence

$$0 \to \mathbb{R} \to \mathbb{R}^m \to \mathbb{R}^m/\mathbb{R} \to 0$$

gives rise to a fibering of $\mathbb{R}^m/H$ over its quotient by H/H_1 with fiber $\mathbb{R}/H_1$. That H/H_1 is discrete in $\mathbb{R}^m/\mathbb{R}$ follows from looking at the images of the $\frac{1}{2}\varepsilon$ balls in $\mathbb{R}^m/\mathbb{R}$. Furthermore, the quotient of $\mathbb{R}^m/\mathbb{R}$ by H/H_1 is compact, connected, and has an $\mathscr{L}(\mathbb{R}^{m-1})$ structure, so the hypothesis of induction applies, and $\mathbb{R}^m/H$ is fibered in circles over the product of $(m-1)$ circles. Also H/H_1 is free abelian of rank $m-1$. This bundle is trivial if and only if its primary obstruction is trivial. Since H has no torsion, the non-triviality of the bundle implies that H is isomorphic to H/H_1, a contradiction. Hence the bundle is trivial and Proposition 3.1 is proved. □

Since there are many compact, connected, parallelizable manifolds which do not admit the structure of an abelian Lie group, the inequivalence of the concepts of bundle reduction and $\mathscr{G}$-structure is proved.

There are many other examples of $\mathscr{L}(\mathbb{R}^m)$ structures. Any open subset of an abelian Lie group has such a structure, because the universal covering group of any abelian Lie group of dimension m is $\mathbb{R}^m$. On the other hand, a proper open subset of $\mathbb{R}^m$ does not inherit the structure of an abelian Lie group, since the closure axiom is not satisfied.

The general problem of finding coordinate systems on a manifold such that the Jacobian matrices of the coordinate changes belong to some specified subgroup G of the general linear group was mentioned in the Preface. It is assumed for this problem that there is given a reduction of the group of the tangent bundle to G, that is, a section in the bundle with fiber $GL_m(\mathbb{R})/G$ associated to the tangent bundle. The condition that the Jacobians belong to G gives a system of partial differential equations for the components of the coordinate transformations. It is necessary to find a complete set of equations, show that they have solutions locally, then find the conditions for existence of global solutions. The integrability problem is precisely this problem of passing from bundle reductions to $\mathscr{G}$-structures. Hence, it is a problem that requires geometric techniques for studying partial differential equations. Some of these will be discussed in § 4.

In the example of $\mathscr{L}(\mathbb{R}^m)$, the section is a frame field $\{v_1, \ldots, v_m\}$. If the frame field is actually tangent to the one parameter subgroups of an abelian Lie group, then the partial differential equations $[v_i, v_j] = 0$ are satisfied. In a coordinate system, $v_i = \partial/\partial x^i$ is a solution, so the local existence problem is trivial. The global problem is nontrivial, and can only be solved in very special cases, as the examples have shown. The assumption of compactness is sufficient but not necessary for the conclusion that an $\mathscr{L}(\mathbb{R}^m)$-structure is an abelian Lie group. The analogous problem arises for many $\mathscr{G}$-structures. A condition necessary to extend a result valid on compact manifolds is known as a completeness condition. These arise in many forms, and it seems necessary to treat each $\mathscr{G}$ separately.

In some cases, integrability can be related to the existence of sufficiently many automorphisms. This question will also be discussed in § 4.

4. Vector Fields and Pfaffian Systems

The first step in solving the integrability problem for a bundle reduction is the local study of the appropriate system of partial differential equations. This section deals with that study for some important special cases, including foliations, complex structures, volume structures, symplectic structures, and contact structures.

In each of these cases, the existence of the bundle reduction is equivalent to the existence locally of a differential form or forms, which are related suitably on overlapping neighborhoods. Instead of looking for a coordinate covering such that the Jacobian matrices of coordinate changes belong to the reduced group, one looks for coordinates in which the defining form or forms have the same expression as in euclidean space. It follows that the coordinate changes belong to the appropriate pseudogroup, so that their Jacobians behave properly. In order for the correct coordinate expressions to exist, the exterior derivatives of the forms may have to satisfy certain conditions. Assuming these conditions, it is necessary to solve some system of partial differential equations to find the coordinate functions. Local solutions have been known for many years, but in some cases only under stringent differentiability conditions, which can now be avoided.

In order to carry out this program in detail, it will be necessary to examine more closely the relations among partial differential equations, differential forms, vector fields, and flows. This examination will be carried out with the case of foliations primarily in mind. Topologically, a submanifold may be viewed locally either as a level set of a submersion or as the image of an immersion. The corresponding infinitesimal notion is tangent plane field, which can be defined either by 1-forms or vector fields. Thus, the plane field spanned by the independent vector fields

$$X_a = \sum_{i=1}^{m} B_a^i(x)\,\partial/\partial x^i \qquad a = 1, \ldots, p$$

is also defined as consisting of vectors annihilating the independent 1-forms

$$\omega^\alpha = \sum_{i=1}^{m} A_i^\alpha(x)\,dx^i \qquad \alpha = 1, \ldots, q$$

where $p + q = m$ and $\sum_{i=1}^{m} A_i^\alpha B_a^i = 0$. The vector fields $\{X_a\}$ generate a pseudogroup, the orbits of which have dimension at least p. Furthermore, the Lie bracket of any two of these vector fields is tangent to the orbit. In particular, a necessary condition for the orbits to have dimension p is that there exist functions f_{ab}^c satisfying

$$[X_a, X_b] = \sum_{c=1}^{p} f_{ab}^c X_c,$$

and this condition will be seen to be sufficient, at least under suitable differentiability hypotheses. Note also that the functions y which satisfy the partial differential equations

$$X_a \cdot y = 0 \qquad a = 1, \ldots, p$$

must be constant on each orbit, whatever its dimension is. From the dual point of view, if there exist submanifolds of dimension p whose tangent spaces annihilate all the ω^α, then each $d\omega^\alpha$ also vanishes on these submanifolds, hence is congruent to zero modulo the ideal generated by $\{\omega^\alpha\}$ in the ring of differential forms. In equations, there exist 1-forms ϕ_β^α such that

$$d\omega^\alpha = \sum_{\beta=1}^{q} \phi_\beta^\alpha \wedge \omega^\beta,$$

a set of conditions which will be seen to be equivalent to the preceding conditions on the brackets. Both sets of conditions have the property that they depend only on the spaces spanned by $\{X_a\}$ or $\{\omega^a\}$, not on the particular choice of bases. A simple example where these conditions are not satisfied is the 2-plane field in $\mathbb{R}^3$ defined as the annihilator of the 1-form

$$\omega = dx^3 + x^2\,dx^1 - x^1\,dx^2.$$

The corresponding pseudogroup is transitive on $\mathbb{R}^3$.

Discussions of the geometric theory of partial differential equations may be found in the books of Forsyth [1890], Kähler [1934], Thomas [1937], and Cartan [1945]. Some interesting historical remarks can be found in the introduction to the paper of Cartan [1899]. Unfortunately, many of the interesting results in the general theory hold only in case of C^ω coefficients. Consequently, the discussion here will consider only those partial differential systems of most interest for the problem of integrability of tangent bundle reductions. In the case of plane fields, the existence and uniqueness of integral manifolds and the structure of the set of accessible points will be discussed, in such a way as to set forth clearly the relationships among the degrees of differentiability of the various objects studied. After that, complex structures, volume structures, symplectic structures, and contact structures will be discussed. Since these are less central to the present work, they will be dealt with more briefly. Since all results are local, it will be assumed that the manifold is an open set in $\mathbb{R}^m$, taken with a fixed C^1 structure for some 1.

Let T be a C^k flow on a C^k manifold for some $k \geqq 1$. Then

$$\frac{d}{dt}(T^i(x,t))|_{t=o} = B^i(x),$$

where $T(x,t) = (T^1(x,t), \ldots, T^m(x,t))$, and

$$X(x) = \sum_{i=1}^{m} B^i(x)\,\partial/\partial x^i$$

is a vector field of class C^{k-1}. Since the concept of C^k vector field makes no sense on a C^k manifold, while a C^{k-1} vector field does not necessarily give rise to a C^k flow, making the concept of flow the basis of the discussion is one way to get the sharpest results. The following lemma about the relation between C^k flows and C^k coordinate systems is the tool necessary to carry through the proofs.

4.1 Lemma. *Let $T(x^1, \ldots, x^m, t)$ be a C^k flow, $k \geqq 1$, generated by a vector field $X(x)$ such that $X(a) \neq 0$. Then there exists a C^k coordinate system $(y^1, \ldots, y^m)$ near a such that $X = \partial/\partial y^1$ and the functions $y^2, \ldots, y^m$ are integrals of T. Hence, a vector field of class C^0 generates at most one C^1 flow. Also, a vector field X of class C^{k-1} generates a C^k flow if and only if there exists a covering by C^k coordinate systems such that X is of class C^k with respect to each system belonging to the covering.*

Proof. The notation introduced before the lemma will be used. By a linear change of coordinates, the condition $B^1(a) \neq 0$ can be obtained, then $T^1(x,t) = a^1$ can be solved for t, say $t = h(x)$ with $h \in C^k$. Let

$$\begin{cases} y^1(x) = -h(x) \\ y^i(x) = T^i(x, h(x)) \quad i > 1. \end{cases}$$

Since for each t, $(\partial T^i/\partial x^j)$ is a nonsingular matrix, and since

$$\frac{\partial h}{\partial x^i} = -\frac{1}{B^1(x)}\frac{\partial T^1}{\partial x^i},$$

the matrix $(\partial y^i/\partial x^j)$ is nonsingular. Note that y^1 is the time parameter on each integral curve, and each other y^i is constant on each integral curve, as required. Consider a C^1 coordinate change that leaves the vector field $\partial/\partial y^1$ fixed. The surface defined by $y^1 = 0$ is mapped into a C^1 submanifold transverse to that vector field, and every such surface is obtained from a unique coordinate change.

Hence, any two C^1 flows arising from a given vector field are related by a diffeomorphism that commutes with the flow, as required. The last statement of the lemma follows easily. □

Next, consider a system of flows $T_a(x, t)$ $a = 1, \ldots, p$ with generators X_a. If the vector fields X_a are tangent to some foliation, then the set accessible from any point is contained in the leaf through that point. Also, for fixed τ, each point accessible by an integral curve of

$$S_t = T_{b,\tau} \circ T_{a,t} \circ T_{b,-\tau} \tag{4.2}$$

is accessible by a curve belonging to the original system. For each point P, the vectors $X_a(P)$ span a subspace of the tangent space at P, and these subspaces form a p-plane field if the dimension is p at each point.

4.3 Definition. The system $\{T_a\}$ is called *complete* if each $T_{b,\tau}$ is an automorphism of the plane field spanned by $\{X_a\}$, that is, the tangent map induced by $T_{b,\tau}$ takes the plane field into itself.

In other words, the system is complete if the generator of each S_t defined by (4.2) is tangent to the plane field.

4.4 Theorem. *Let $\{T_a \mid a = 1, \ldots, p\}$ be a complete system of C^k flows on a C^l manifold, $l \geqq k \geqq 1$, such that the generators X_a are independent. Then any point has a neighborhood U on which are defined C^k functions $\{y^\alpha \mid \alpha = 1, \ldots, q = n - p\}$ such that*

(i) *Each y^α is an integral for each T_a.*
(ii) *The forms dy^α are independent.*
(iii) *The subset of U accessible from any point is ptrecisely the leaf through that point of the foliation defined by the functions y^α.*

Proof. This theorem is proved by Chow [1940/41]. The proof uses an induction, as do most older proofs related to the integrability of plane fields. But the formulation in terms of flows avoids the loss of differentiability that restricts most other proofs to the C^∞ case. The idea of the proof is to change coordinates so that one of the flows is generated by $\partial/\partial x^1$, then examine the remaining flows to show that the hypothesis of induction can be applied. □

An immediate consequence of this theorem is that if a complete system $\{T_a\}$ is given on any manifold such that the dimension of the set of vectors $\{X_a(P)\}$ is independent of P, then the accessible sets form a C^k foliation of the manifold. The necessity of the condition of constant dimension is shown by an example of Chow [1940/41]. The discussion of incomplete systems falls outside foliation theory, and will be taken up briefly in § 6. For the discussion of foliations from the point of view of forms and vector fields, it will be convenient to consider a C^{k+1} manifold with C^k forms and vector fields, still for $k \geqq 1$. The first step is to determine the conditions on a vector field in order that the corresponding flow consist of automorphisms of some given plane field.

4.5 Lemma. *Let a C^k p-plane field be spanned by C^k vector fields Y_a, $a = 1, \ldots, p$, and let its annihilating cotangent field be spanned by C^k forms ω^α, $\alpha = 1, \ldots, q$. Let X be a C^k vector field. Then the following conditions are equivalent:*

(i) *The flow T_t generated by X consists of automorphisms of the plane field.*

(ii) *For every a, there exist C^{k-1} functions g_a^b such that*

$$\mathscr{L}_X Y_a = [X, Y_a] = \sum_{b=1}^{p} g_a^b Y_b.$$

(iii) *For every α, there exist C^{k-1} functions f_β^α such that*

$$\mathscr{L}_X \omega^\alpha = \sum_{\beta=1}^{q} f_\beta^\alpha \omega^\beta.$$

Proof. Suppose (iii) holds. On the one hand

$$(\mathscr{L}_X \omega^\alpha)(Y_a) = \sum_{\beta=1}^{q} f_\beta^\alpha \omega^\beta (Y_a) = 0.$$

On the other hand

$$\begin{aligned}(\mathscr{L}_X \omega^\alpha)(Y_a) &= ((i_X d + d i_X)\omega^\alpha)(Y_a) \\ &= d\omega^\alpha(X, Y_a) + Y_a \omega(X) \\ &= X \cdot \omega^\alpha(Y_a) - \omega^\alpha([X, Y_a]) = -\,\omega^\alpha([X, Y_a]).\end{aligned}$$

Since $[X, Y_a]$ is annihilated by all ω^α, and since $\mathscr{L}_X Y_a = [X, Y_a]$ for any pair of vector fields, condition (ii) holds.

Next suppose (ii) holds. By definition,

$$\begin{aligned}\mathscr{L}_X Y_a &= \lim_{t\to 0} \frac{T^*_{-t}(Y_a) - Y_a}{t} \\ &= \frac{d}{dt}(T^*_{-t}(Y_a))_{t=0}.\end{aligned}$$

In the last equation, $T^*_{-t}(Y_a)$ is to be considered as a vector field on $\mathbb{R}^m \times \mathbb{R}$. Hence

$$\begin{aligned}0 = \omega^\alpha(\mathscr{L}_X Y_a) &= \lim_{t\to 0} \frac{1}{t}\{\omega^\alpha(T^*_{-t}(Y_a)) - \omega^\alpha(Y_a)\} \\ &= \frac{d}{dt}(\omega^\alpha(T^*_{-t}(Y_a)))_{t=0}.\end{aligned}$$

This implies that the function

$$F(t) = \omega^\alpha(T^*_{-t}(Y_a))$$

satisfies $F(0) = 0$ and $F'(t) = 0$ for all t. Hence condition (i) holds.

Finally, suppose (i) holds. Then from

$$\mathscr{L}_X \omega^\alpha = \lim_{t\to 0} \frac{1}{t}\{T_t^* \omega^\alpha - \omega^\alpha\}$$

and the fact $T_t^* \omega^\alpha$ is a linear combination of the ω^β, it follows that $\mathscr{L}_X \omega^\alpha$ has the same property. Hence (i) implies (iii) and the lemma is proved. □

4.6 Definition. A vector field which satisfies the conditions of Lemma 4.5 is called an *infinitesimal automorphism* of the plane field.

Each of the following two propositions gives a set of equivalent conditions, but for finite differentiability the two sets are not equivalent to each other. The exact relation will be formulated afterwards as Theorem 4.11. The desirability of obtaining this precision is indicated by the results of Harrison [1975, 1979], according to which on any manifold of dimension at least two there exist C^k diffeomorphisms which are not topologically conjugate to any C^{k+1} diffeomorphism.

4.7^k Proposition. *Let a C^k plane field be given on a manifold of class C^l, $l > k \geq 1$. Then if a point P has a neighborhood on which one of the following conditions is satisfied, it has a neighborhood on which all are satisfied:*

(i) *Every C^k vector field tangent to the plane field generates a flow consisting of automorphisms of the field.*

(ii) *For any C^k basis $\{X_a\}$ for the plane field, there exist C^{k-1} functions f^c_{ab} such that*

$$\mathscr{L}_{X_a} X_b = [X_a, X_b] = \sum_{c=1}^{p} f^c_{ab} X_c$$

(iii) *For any C^k basis $\{X_a\}$ for the plane field and any C^k basis $\{\omega^\alpha\}$ for the annihilating coplane field, there exist C^{k-1} functions $g^\alpha_{a\beta}$ such that*

$$\mathscr{L}_{X_a} \omega^\alpha = \sum_{\beta=1}^{q} g^\alpha_{a\beta}\, \omega^\beta$$

(iv) *For any C^k basis $\{\omega^\alpha\}$ for the annihilating coplane field, there exist C^{k-1} forms ϕ^α_β such that $d\omega^\alpha = \sum_{\beta=1}^{q} \phi^\alpha_\beta \wedge \omega^\beta$.*

(v) *For any C^k basis $\{\omega^\alpha\}$ for the annihilating coplane field, let $\Omega = \omega^1 \wedge \ldots \wedge \omega^q$. Then for every α, $d\omega^\alpha \wedge \Omega = 0$.*

(vi) *There exists a C^k basis $\{Z_a\}$ for the plane field such that $[Z_a, Z_b] = 0$.*

Proof. The first three conditions are equivalent by Lemma 4.5. The fourth and fifth are equivalent by a theorem of exterior algebra.

Since $\omega^\alpha(X_a) = 0$,

$$d\omega^\alpha(X_a, X_b) = -\,\omega^\alpha([X_a, X_b]). \tag{4.8}$$

On the other hand, condition (iv) implies

$$d\omega^\alpha(X_a, X_b) = 0 \tag{4.9}$$

which together with (4.8) implies (ii). Conversely, (ii) and (4.8) together imply (4.9). Let ξ^a be forms such that $\{\xi^a, \omega^\alpha \mid a = 1, \ldots, p,\ \alpha = 1, \ldots, q\}$ is a basis for the cotangent space. Then by (4.9)

$$d\omega^\alpha = \sum_{a,\beta} A^\alpha_{a\beta}\, \xi^a \wedge \omega^\beta + \sum_{\beta,\gamma} B^\alpha_{\gamma\beta}\, \omega^\gamma \wedge \omega^\beta$$

so that taking $\phi^\alpha_\beta = \sum_a A^\alpha_{a\beta}\, \xi^a + \sum B^\alpha_{\gamma\beta}\, \omega^\gamma$ satisfies condition (iv). Hence, the second and fourth conditions are equivalent.

Choose an arbitrary C^{k+1} coordinate system, and write the vectors X_a in condition (ii) in terms of the coordinate basis:

$$X_a = \sum_i A_a^i \, \partial/\partial x^i.$$

Since the matrix with entries A_a^i has rank p, multiplying by the inverse of a nonsingular $p \times p$ submatrix and permuting the coordinates gives

$$Z_a = \partial/\partial x^a + \sum_\alpha A_a^\alpha \frac{\partial}{\partial x^{p+\alpha}}$$

as a new basis for the same plane field. Then $[Z_a, Z_b]$ satisfies condition (ii) if and only if $[Z_a, Z_b] = 0$. On the other hand, if one basis for the plane field satisfies condition (ii), they all do. Hence, (ii) is equivalent to (vi) and Proposition 4.7^k is proved. □

4.10^k Proposition. *Let a C^{k-1} plane field be given on a manifold of class C^l, $l \geqq k \geqq 1$. If $k \geqq 2$, then if a point P has a neighborhood on which one of the following four conditions is satisfied, it has a neighborhood on which they all are satisfied. If $k = 1$, then the same conclusion holds, except that only the first two conditions hold.*

(i) *There is a complete system $\{T_a | a = 1, \ldots, p\}$ of C^k flows whose generators $\{X_a\}$ span the p-plane field.*

(ii) *There exist C^k functions $\{y^\alpha | \alpha = 1, \ldots, q\}$ such that $\{dy^\alpha\}$ is a basis for the annihilating coplane field.*

(iii) *There exist C^{k-1} forms ω^α which are closed and form a basis for the annihilating coplane fields.*

(iv) *Each C^{k-1} basis $\{Z_a | a = 1, \ldots, p\}$ for the plane field can be extended by a C^k basis $\{N_\alpha | \alpha = 1, \ldots, q\}$ for a complementary plane field so that the commutator of any two elements of the basis $\{Z_a, N_\alpha\}$ for the tangent space to the manifold is tangent to the plane field.*

Proof. The first condition implies the second by Theorem 4.4. Assume the second and choose C^k functions $\{x^a | a = 1, \ldots, p\}$ such that $(x^1, \ldots, x^p, y^1, \ldots, y^q)$ form a coordinate system. Then the flows T_a defined by

$$T_a(x, y, t) = (x^1, \ldots, x^{a-1}, x^a + t, x^{a+1}, \ldots, x^p, y^1, \ldots, y^q)$$

satisfy the first condition. The second implies the third, since one may take $\omega^\alpha = dy^\alpha$. Conversely, given $\{\omega^\alpha\}$, choose any C^k coordinates $(x^1, \ldots, x^n)$ and write

$$\omega^\alpha = \sum_{i=1}^n \omega_i^\alpha(x)\, dx^i$$

Let $y^\alpha = \int_0^1 \left(\sum_{i=1}^n x^i \omega_i^\alpha(tx) \right) dt$. Then $dy^\alpha = \omega^\alpha$, so that $\{y^\alpha\}$ satisfies condition (ii).

The proof that the third and fourth conditions are equivalent is based on the observation that if $\omega^\alpha(Y_i)$ is constant, then

$$d\omega^\alpha(Y_i, Y_j) = -\,\omega^\alpha([Y_i, Y_j]).$$

Assuming condition (iii), choose any C^{k-1} basis $\{Z_a\}$ for the plane field, and any vector fields N_α such that $\omega^\alpha(N_\beta) = \delta_\beta^\alpha$. Then these satisfy condition (iv). Conversely, given $\{Z_a, N_\alpha\}$ satisfying (iv), define ω^α by $\omega^\alpha(Z_a) = 0$ and $\omega^\alpha(N_\beta) = \delta^\alpha_\beta$. This concludes the proof of the proposition. □

The fourth condition of Proposition 4.10^k has an immediate interpretation in terms of automorphisms: Any basis for the plane field can be extended to a transitive family of automorphisms which is commutative modulo the plane field. This formulation admits a C^0 version which will be discussed later. Meanwhile, it is now possible to formulate and prove the precise relation between plane fields and foliations, including the relation between the conditions of Propositions 4.7^k and 4.10^k.

4.11 Theorem. *For all $k \geqq 1$, the conditions of Proposition 4.10^{k+1} imply those of Proposition 4.7^k, the conditions of Proposition 4.7^k imply those of Proposition 4.10^k, and a $\mathscr{G}^k_{p;\,m}$-structure satisfies the conditions of Proposition 4.10^k.*

Proof. Assume Proposition 4.10^{k+1} holds, and let $\{\omega^\alpha\}$ be the basis mentioned in condition (iii). Then $d\omega^\alpha = \sum_{\beta=1}^{q} \phi^\alpha_\beta \wedge \omega^\beta$ where $\phi^\alpha_\beta = 0$, so condition (iv) of Proposition 4.7^k is satisfied. Conversely, condition (vi) of Proposition 4.7^k implies that the C^k flows T_a generated by the Z_a form a complete system, so that condition (i) of Proposition 4.10^k is satisfied. Condition (ii) of Proposition 4.10^k is equivalent to the existence of a $\mathscr{G}^k_{p;\,m}$-structure. □

The difference between the conditions in the two propositions reflects the fact that on a C^k manifold, the linear tangent bundle is a C^{k-1} manifold. An alternative approach which is useful also for topological manifolds, is based on the fact that the group of C^k diffeomorphisms of $\mathbb{R}^n$ is a topological group $\mathrm{Diff}^k(\mathbb{R}^n)$ in the C^k topology, $k \geqq 0$. Hence, the tangent bundle of a C^k manifold M has the structure of a bundle with structure group the group of C^k diffeomorphisms of $\mathbb{R}^m$ leaving the origin fixed. (The condition on the origin is possible because the quotient space of the full group by the subgroup leaving the origin fixed is $\mathbb{R}^m$, hence is contractible.) The total space of this tangent bundle is C^k equivalent to a neighborhood of the diagonal in $M \times M$, with the projection induced by projection on the first factor. A p-plane field is the reduction of the group of this bundle to the subgroup consisting of the diffeomorphisms taking a given p-dimensional linear subspace onto itself. Since taking the Jacobian at the origin defines a continuous homomorphism of $\mathrm{Diff}^k(\mathbb{R}^m)$ onto the general linear group, a p-plane field in this sense defines a p-plane field in the sense of the linear tangent bundle. Moreover, any C^k diffeomorphism f induces a diffeomorphism of $M \times M$, hence a mapping f_* taking a p-plane field onto a p-plane field. Thus, the notion of C^k automorphism of such a field makes sense, namely a C^k diffeomorphism f such that f_* takes the field onto itself. A foliation of M gives rise to a foliation of each fiber in the total space of the tangent bundle, viewed as a neighborhood of the diagonal in $M \times M$, and therefore to a p-plane field. An integrability criterion for p-plane fields is given by the following proposition.

4.12^k Proposition. *Let $k \geqq 0$, and let a C^k plane field be given on a C^k manifold. Suppose each point has a neighborhood C^k diffeomorphic to $\mathbb{R}^n$ on which the plane* *field admits a transitive abelian group of C^k automorphisms. Then there is a C^k foliation which gives rise to this plane field.*

Proof. See Reinhart [1971]. The proof relies heavily upon the fact that up to isomorphism in the category of topological groups, the additive group of $\mathbb{R}^m$ is the only simply connected abelian topological group which is a topological manifold of dimension m. The only fact about plane fields required is that a plane field on $\mathbb{R}^m$ which is invariant by all translations is necessarily integrable. Thus, the proposition extends to many other interesting pseudogroup structures. Note that since there is no loss of differentiability, this result applies to topological manifolds as well. □

In summary, integrability means translation invariance. The complications in the various formulations arise from the fact that the preceding statement must be made in a properly chosen coordinate system.

This completes the local existence theory for foliations. Much of the global theory belongs to topology rahter than differential geometry, but some aspects will be discussed in Chapter III.

The integrability theorem for plane fields, or rather the corresponding theorem in partial differential equations, was first proved by Deahna [1840], and then by Clebsch [1866]. Frobenius [1877] was the first to use something like the exterior derivative in his formulation. For applications in thermodynamics, Cartheodory [1909] proved that integrability of a nonzero 1-form is equivalent to the existence of inaccessible points near any given point. A generalization to Pfaffian systems was found by Chow [1940/41]. The proof of the Deahna-Clebsch-Frobenius theorem given here is essentially a commentary on Chow's theorem, since the hard part is the proof of Theorem 4.4, which is entirely due to Chow.

The next integrability problem to be discussed will be that for complex structures. In this discussion, it is convenient to consider complex valued functions, differential forms, and vector fields on $\mathbb{R}^m$, where $m = 2n$. (Abstractly, such forms or vector fields are sections of the tensor product of the real cotangent or tangent bundle with $\mathbb{C}$.) Such an object is C^k if its real and imaginary parts are each separately C^k. A reduction of the group of the tangent bundle on a contractible open set in $\mathbb{R}^{2n}$ to the complex linear group is equivalent to giving on this set a family of complex valued vector fields

$$X_\alpha = \sum_{i=1}^{2n} a_\alpha^i(x) \frac{\partial}{\partial x^i}$$

such that $\{X_\alpha, \bar{X}_\alpha \mid \alpha = 1, \ldots, n\}$ is a basis for the complex tangent vectors at each point of the set. Here $\bar{X}_\alpha = \sum_{i=1}^{2n} \bar{a}_\alpha^i(x)\, \partial/\partial x^i$ and $\bar{a}$ denotes the complex conjugate of a. Two such bases define the same almost complex structure if they are related by a nonsingular n-dimensional complex linear transformation. Given $\{X_\alpha\}$, a basis $\{\omega^\alpha, \bar{\omega}^\alpha\}$ for the complex cotangent space is obtained by setting

$$\omega^\alpha(X_\beta) = \delta_\beta^\alpha \qquad \omega^\alpha(\bar{X}_\beta) = 0.$$

Alternatively, one may give $\{\omega^\alpha\}$ first and obtain $\{X_\alpha\}$ from them.

4.13 Definition. A function on $\mathbb{R}^m$ is of *class* $C^{k+\lambda}$, $0 < \lambda < 1$, if it is of class C^k and its derivatives of order k satisfy inequalities of the form

$$|f(x) - f(y)| \leqq C x - y^\lambda$$

uniformly on compact subsets of the domain of the function.

4.14 Theorem. *Suppose an almost complex structure on* $\mathbb{R}^m$ *is of class* $C^{k+\lambda}$, *where* $k \geqq 1$ *and* $0 < \lambda < 1$, *and is defined by differential forms* $\{\omega^\alpha\}$ *satisfying* $d\omega^\alpha = \sum \phi^\alpha_\beta \wedge \omega^\beta$ *for some 1-forms* $\{\phi^\alpha_\beta\}$. *Then in the neighborhood of any point there are complex valued functions* $\{z^\alpha\}$ *of class* $C^{k+1+\lambda/n}$ *which define a complex structure associated to the given almost complex structure.*

Despite the similarity between the statements of the integrability condition for plane fields and almost complex structures, the differential equations to be solved are of totally different types. For the almost complex case, the equations are elliptic, and the first solution for $n > 1$ and data which are not real analytic is due to Newlander and Nirenberg [1957]. Their method was refined by Nijenhuis and Woolf [1963] to give the result stated here. Note that for $k = 1$, the integrability conditions stated here do not make sense, and a suitable variation must be introduced, Also, for $n = 1$, the integrability condition is automatically satisfied, and the theorem has been known for a long time. Nonetheless, a new proof by Chern [1955] for $k \geqq 0$ was important in preparing the way for the higher dimensional result.

There is an interesting theorem which can be obtained by combining the results on foliations and on complex structures, but some preliminary remarks will be required. Given a real vector space V and a subspace C of $V \otimes \mathbb{C}$, there are subspaces A and B of V such that

$$C + \bar{C} = A \otimes \mathbb{C}$$
$$C \cap \bar{C} = B \otimes \mathbb{C}.$$

If a field of subspaces C_x of $T^*\mathbb{R}^m \otimes \mathbb{C}$ has a local basis $\{\omega^\alpha\}$ satisfying the integrability condition $d\omega^\alpha = \sum \phi^\alpha_\beta \wedge \omega^\beta$, then the corresponding field A_x is integrable, but B_x need not be. Furthermore, if the dimensions of B_x and C_x are independent of x, so is the dimension of A_x. These facts show that the hypotheses of the following proposition are reasonable.

4.15 Proposition. *In an open set of* $\mathbb{R}^m$, let $\{\omega^A\}$ *be* C^k *complex valued 1-forms which are a basis for a field* C_x *of subspaces of* $T^* \otimes \mathbb{C}$ *and satisfy* $d\omega^A = \sum \phi^A_B \wedge \omega^B$ *for some 1-forms* ϕ^A_B *of class* C^{k-1}. *Suppose further* $\{\theta^a\}$ *are a* C^k *basis for* $C_x \cap \bar{C}_x$ *and satisfy* $d\theta^a = \sum \psi^a_b \wedge \theta^b$ *for some* ψ^a_b *of class* C^{k-1}. *Let* $k \geqq 3$. *Then, in the neighborhood of any point there are* C^k *coordinates* $(x^i, z^\alpha, \bar{z}^\alpha, y^a)$ *such that* $\{dy^a\}$ *define the same foliation as* $\{\theta^a\}$, $\{dz^\alpha, d\bar{z}^\alpha, dy^a\}$ *define the same foliation as* $\{\omega^A, \bar{\omega}^A\}$, *and* $\{z^\alpha\}$ *define a complex structure on the local quotient of the higher dimensional foliation by the lower dimensional foliation.*

Proof. Choose C^k functions (y^a) that define the same foliation as $\{\theta^a\}$ and C^k functions $\{\xi^A\}$ that define the same foliation as $\{\omega^A, \bar{\omega}^A\}$. Select a subset $\{\xi^\alpha\}$ of $\{\xi^A\}$ so that (ξ^α, y^a) also define the latter foliation. By adding C^k functions x^i, one obtains a C^k coordinate system (x^i, ξ^α, y^a) in which both real foliations are C^k and the ω^α are C^{k-1}. The restrictions of the ω^α to a cross-section defined by holding the x^i constant gives integrable almost complex structures on each leaf of the foliation defined by $\{dy^a\}$, and these structures are C^{k-1} functions of the parameters (y^a). Since Nijenhuis and Woolf (in the paper quoted above) prove a parameterized version of their theorem, functions $z^\alpha(\xi^\alpha, y^a)$ exist with all the required properties. □

A proposition of this type was first proved by Nirenberg [1957], and the preceding proof is essentially his. Hörmander [1964] gives a direct proof, using the dual formulation in terms of vector fields.

There remain the problems of finding local normal forms for volume, symplectic, and contact structures. These three cases can be treated in very similar ways. In each case, the structure is reduced to normal form at one point by a linear change of coordinates, then the coordinate system is deformed so that the given structure takes this normal form on a neighborhood of the point. The deformation is found by integrating a time dependent vector field, chosen by examining the condition that the time derivative of the structure be zero. This condition derives from the following lemma. In the statement of the lemma, differentiability conditions apply jointly to the variables (x, t), but terms like diffeomorphism, vector field, and form apply only to the variable x. Thus, for example, a vector field $v_t(x) = v(x, t)$ is a function of (x, t) whose values are vectors tangent to x-space.

4.16 Lemma. *Let f_t be a C^k family of diffeomorphisms defined on a neighborhood of the origin in $\mathbb{R}^m$ such that for some C^k vector field v_t*

$$\frac{df_t}{dt}(x) = v_t(f_t(x)).$$

Further, suppose ϕ_t, $0 \leqq t \leqq 1$, is a C^k family of differential forms defined near the origin, and $k \geqq 1$. Then

$$\frac{d}{dt}[f_t^* \phi_t] = f_t^* \left[\frac{d\phi_t}{dt} + \mathscr{L}_{v_t} \phi_t\right].$$

Proof. The x variables in the expression $f_t^* \phi_t$ depend on t by means of the vector field. In addition, ϕ_t is a function of t. Hence

$$\frac{d}{dt}[f_t^* \phi_t]_{t=s} = \frac{d}{dt}[f_t^* \phi_s + f_s^* \phi_t]_{t=s}.$$

On the other hand, by a formula which can be found in Koboyoshi and Nomizu [1963, p. 32]

$$\frac{d}{dt}[f_t^* \phi_s]_{t=s} = f_s^* (\mathscr{L}_{v_s} \phi_s).$$

This completes the proof of the lemma. □

An argument using a time-dependent vector field was first applied by Moser [1965] to show that the only invariant for a volume element on a compact manifold is the total volume. A modification was used by Weinstein [1977, Lecture 5] to show that by a suitable choice of local coordinates, every C^1 volume form can be reduced to $dx^1 \wedge \ldots \wedge dx^m$. The same method shows that locally every C^1 closed 2-form of maximal rank can be reduced to $\sum_{i=1}^{n} dx^i \wedge dx^{i+n}$ and every C^2 1-form of maximal rank can be reduced to $\mathrm{d}x^{2n+1} + \sum_{i=1}^{n} x^i \mathrm{d}x^{i+n}$. Note that in the volume and contact cases there is no integrability condition, while in the symplectic case the integrability condition is that the 2-form be closed.

This completes the discussion of the local theory of integrability for bundle reductions. The global problem is to find bundle reductions which satisfy the integrability conditions, and classify them in some way. In the case of foliations, a solution exists, and will be discussed briefly in Chapter III. For most other structures, very little is known, but it is to be hoped that some of the methods developed to study foliations can be generalized.

5. Leaves and Holonomy

The last few sections have dealt with the most essential local properties of foliations and other pseudogroup structures. In this section, the first steps are taken toward elucidating the global structure of foliations. Many of the concepts introduced will be generalizations of notions in topological dynamics – the correspondence can be seen by replacing "leaf" by "orbit of a group action" and making other changes that will be indicated below.

5.1 Definition. A *saturated set* (or an *invariant set*) on a foliated manifold is a subset which is the union of leaves. A *minimal set* is a closed saturated set which has no proper closed saturated subsets.

Any compact saturated set contains at least one minimal set. The proof, which is based on Zorn's lemma, is the same as the proof in topological dynamics. A minimal set is the closure of each leaf contained in it, since if not, the closure would be a proper closed saturated subset of the minimal set. If the minimal set consists of more than one leaf, any transverse submanifold of dimension equal to the codimension of the foliation (compact or not, but without boundary), meets either every leaf belonging to the minimal set, or none of them. The intersection is closed in the transverse submanifold and dense in itself, hence uncountable, so that the number of leaves is also uncountable. A minimal set with a nonvoid interior is open. The proofs of all these statements are essentially the same as those for the corresponding statements in topological dynamics, and will be omitted.

For a geometric understanding of foliations, it is necessary to see how a leaf or a minimal set is related to nearby leaves. Restricted to a small enough neighborhood the leaves are parallel, but very complicated situations can arise when a leaf or minimal set is looked at globally. It is convenient to introduce two classes of examples in order to get an idea of the possibilities.

5.2 Example. If $f: M \to M$ is a C^k diffeomorphism, let V be the quotient of $M \times [0,1]$ obtained by identifying $(x,0)$ with $(f(x),1)$. V has a unique 1-dimensional C^k foliation induced by the foliation of $M \times [0,1]$ with leaves $\{x\} \times [0,1]$ and called the suspension of f. The suspension has a cross-section induced by the subset $M \times \{0\}$. f, regarded as a map of this cross-section, determines the foliation completely. f may be regarded as the generator of a cyclic group, so that the topological dynamics of this group and that of the foliation are simply two viewpoints on the same phenomenon. The set of germs of powers of f is a groupoid which also contains all the dynamical information, as does the pseudogroup of restrictions of f to open sets. If x is a fixed point f, then $\{x\} \times [0,1]$ determines a compact leaf of the suspension, and the restriction of f to any neighborhood of x is a diffeomorphism onto some other neighborhood of x. If x is a periodic point it also gives rise to such a local diffeomorphism. These diffeomorphisms, viewed as mappings of a local cross-section, have been studied in the theory of ordinary differential equations (under the name of "Poincaré map") because they determine the behaviour of solutions near to any given periodic orbit.

5.3 Example. Any smooth fiber bundle can be viewed as a bundle with group the full group of diffeomorphisms of the fiber. If it is then equivalent to a bundle in which this group has the discrete topology, there is foliation transverse to the fibers and of complementary dimension, constructed as in Example 2.10. By the same construction, one obtains a homomorphism h of the fundamental group of the base space into the diffeomorphism group of the fiber which determines the bundle. Note that Example 5.2 is precisely the special case in which the base space is S^1. In general, each leaf of the transverse foliation is a covering space of the base, and its fundamental group at any point x is isomorphic to the set of elements α of the fundamental group of the base space such that $h(\alpha)$ leaves x fixed. Either $h(\alpha)$ or its restriction to a neighborhood of x may be regarded as a "Poincaré map" associated to α. Since in general the fundamental group of the base can be very complicated, the structure of the set of "Poincaré maps" can be very complicated.

Note that the existence of a foliation of complementary dimension transverse to the fibers is not sufficient to imply that the group of a bundle is discrete. To see what is true, immerse a circle in the base space and pull back the given bundle to a bundle over the circle. This latter bundle has a 1-dimensional foliation transverse to the fibers, so that the tangent vector field to the circle determines a vector field on the total space. This vector field can be integrated to give 1-parameter flows defined on small open sets for finite time. If the fiber is compact, then the flow is defined for all time at every point. Since the fundamental group of the base space of the original bundle is generated by immersed circles, this construction can be applied to a generating system to show that the group of the bundle is discrete.

Example 5.3 shows that for an arbitrary foliation, it is desirable to study mappings of cross-sections associated to curves lying in leaves. In the case of a single leaf or a minimal set, it is sufficient to consider mappings whose domain is contained in a fixed transverse disk. The case of a single leaf will be dealt with first. Two quite different constructions will be given, one more pictorial, the other more abstract but giving sharper differentiability results.

5.4 Proposition. *Let L be a leaf of a foliation of codimension q, P a point of L, and h an embedding of $\mathbb{R}^q$ transverse to the foliation such that $h(0) = P$. If the foliation is of class C^k, then h is assumed to be of class C^k also. Then there is a homomorphism $f(L, P, h)$ taking $\pi_1(L, P)$ into the group of germs of C^k diffeomorphisms of $\mathbb{R}^q$ leaving 0 fixed. If $f(L, Q, h_1)$ is another such homomorphism and α is a homotopy class of paths from P to Q, then*

$$f(L, Q, h_1)(\alpha^{-1}\gamma\alpha) = h_1^{-1} \circ h \circ f(L, P, h)(\gamma) \circ h^{-1} \circ h_1.$$

In brief, f is determined up to conjugacy by the foliation and the leaf.

First proof. It was shown in § 2 that the normal bundle to any leaf admits a foliation transverse to the fibers having the zero section as a leaf. Let γ be an element of $\pi_1(L, P)$, and represent γ by a map g: $[0, 1] \to L$, such that $g(0) = g(1) = P$. Then the bundle $g^*(v(L))$ admits a 1-dimensional foliation transverse to the fibers, and h induces an embedding $\tilde{h}$ of a neighborhood of 0 in $\mathbb{R}^q$ into the total space of the bundle. If π is the projection in this bundle, then $\pi^{-1} \circ g$ makes each leaf into a path, defined on some open subinterval of $[0, 1]$. Since the zero section is a leaf, paths having $(\pi^{-1} \circ g)(0)$ close to 0 will be defined an all of $[0, 1]$. The map taking the initial point of each such path to its final point is a diffeomorphism of some neighborhood of 0 in the fiber to some other neighborhood of 0. Though the image of $\tilde{h}$ does not necessarily lie in this fiber, it still induces a uniquely defined germ of a diffeomorphism at the origin in $\mathbb{R}^q$. This germ is taken to be $f(L, P, h)(\gamma)$. The verification that all necessary properties hold is tedious but straightforward. Furthermore, in case of finite differentiability, the use of the linear tangent bundle makes it impossible to obtain f of class C^k. For these reasons, the rest of the details of the first proof will be omitted. However, this construction should not be ignored, since it is sometimes useful. □

Second proof. For each point P on the foliated manifold M, consider the set of germs at P of local submersions onto $\mathbb{R}^q$, as defined in § 2. The set of all such germs at all P forms a set $\tilde{M}$ which admits a natural projection π onto M by taking each germ into the point at which it is defined. $\tilde{M}$ is given the smallest topology such that each set of the following form is open: the set of all germs of a given local submersion at all points of some open set contained in its domain. Since π is then a local homeomorphism, the leaf topology may be lifted from M to a leaf topology in $\tilde{M}$ so that π is also a local homeomorphism of the leaf topologies.

5.5 Lemma. *π: $\tilde{M} \to M$ is a covering when the spaces are given the leaf topologies.*

Proof. See Haefliger [1958, 1962]. □

Given h and P as in the hypotheses of the Proposition, there is a unique $\tilde{P}$ which is a germ of a submersion at P such that $\tilde{P} \circ \tilde{h}$ is the germ of the identity at 0, where $\tilde{h}$ is the germ of h at 0. Since π is a covering, any path starting at P admits a unique lift starting at $\tilde{P}$, whose endpoint depends only on the homotopy class (with fixed endpoints) of the path. For a homotopy class γ of closed paths, this endpoint is another germ Q of a submersion that takes P to 0. Then $f(L, P, h)(\gamma)$ is the germ g of a homeomorphism of $\mathbb{R}^q$ which is uniquely defined by the condition that $Q = g \circ \tilde{P}$. Since g is unique, it is easy to verify that all the other conditions of the Proposition are satisfied. □

In the case of a compact leaf, Haefliger [1958] has shown that if $k > 0$, the foliated structure in the neighborhood of the leaf is determined up to diffeomorphism by the holonomy homomorphism. More generally, this is true if the leaf is proper, in the sense that its leaf topology is the same as the topology induced on it as a subspace of M.

5.6 Definition. The *holonomy homomorphism* of a leaf is the homomorphism occurring in the statement of Proposition 5.4. The *holonomy group* of a leaf is the image of the holonomy homomorphism, and the *holonomy covering* is the covering whose fundamental group is the kernel of the holonomy homomorphism.

In the case of Example 5.3, the fundamental group of a leaf can be identified with a subgroup of the fundamental group of the base, in such a way that the holonomy homomorphism can be computed from a knowledge of the structural homomorphism of the bundle.

In the case of an improper leaf, the holonomy group does not contain all the information about the structure of leaf neighborhoods, because there is interesting information associated to curves that are very close to being closed in the manifold topology, but not in the leaf topology. Indeed, each homotopy class with fixed endpoints of curves lying on a leaf has associated with it various diffeomorphisms between local quotients at the endpoints. The method used in the first proof of Proposition 5.4 can be used to define a holonomy pseudogroup associated to the set of all such homotopy classes of not necessarily closed curves. Passing to germs gives the holonomy groupoid, which can also be obtained by generalizing the second construction. This groupoid is also known as the graph of the foliation. It will be discussed in detail in § IV.2.

One important kind of cross-section consists of the disjoint union of countably many disks. Such a cross-section can be obtained from a countable covering by coordinate neighborhoods, and thus have the additional property that it meets each leaf. It is useful to inquire how the holonomy group of the leaves can be constructed from the elements of the holonomy pseudogroup of such a cross-section. Associated to a closed curve on a leaf is a fixed point of some element of the pseudogroup, but the corresponding element of the holonomy group is trivial unless this fixed point lies in the boundary of the set of fixed points. But the boundary of a set is a closed, nowhere dense set. This is the idea behind the following result of Epstein, Millett, and Tischler [1977], which was also obtained by Hector [1977].

5.7 Proposition. *The union of all leaves with trivial holonomy is a dense set which is the intersection of countably many open sets.*

Proof. The idea of the proof has been given above. The essential technical tool is the choice of a covering by foliation coordinate systems of the type whose existence is guaranteed by Proposition 2.5. For details, see the papers cited above. ▯

Associated with any holonomy group (or groupoid or pseudogroup) is a family of infinitesimal holonomy groups (or groupoids), obtained by taking the j-jets, $j \leqq k$, of the elements of the holonomy group (or groupoid or pseudogroup). The most interesting member of this family is the linear holonomy, that is, the case $j = 1$. For any closed curve, this is an element of the general linear group. There is a useful integral formula for the determinant of the linear holonomy along a curve. The integrand is a differential 1-form constructed out of the forms which define the foliation locally, as follows:

Let the forms ω^α such that

$$d\omega^\alpha = \sum \omega^\alpha_\beta \wedge \omega^\beta$$

define the foliation on some open set, and let

$$\Omega = \omega^1 \wedge \ldots \wedge \omega^q.$$

Then

$$d\Omega = \operatorname{tr}(\omega^\alpha_\beta) \wedge \Omega$$

where

$$\operatorname{tr}(\omega^\alpha_\beta) = \sum_\alpha \omega^\alpha_\alpha.$$

Furthermore, the restriction of $\operatorname{tr}(\omega^\alpha_\beta)$ to any leaf is closed, since

$$d(\operatorname{tr}(\omega^\alpha_\beta)) \wedge \Omega = 0.$$

If the foliation is defined also by ϕ^α, where $\phi^\alpha = A^\alpha_\beta \omega^\beta$, and if ϕ^α_β and Φ are defined as above, then

$$\operatorname{tr}(\phi^\alpha_\beta) \wedge \Omega = [d(\ln|\det A^\alpha_\beta|) + \operatorname{tr}(\omega^\alpha_\beta)] \wedge \Omega$$

so that on a leaf, the two traces differ by an exact form. Thus, the trace gives rise to a well-defined cohomology class on each leaf.

5.8 Proposition. *The integral along any C^1 closed curve γ lying on a leaf of the 1-form* $\operatorname{tr}(\omega^\alpha_\beta)$ *is equal to the natural logarithm of the absolute value of the determinant of the linear holonomy map along γ.*

Proof. See Reeb [1952, p. 116]. ▯

In codimension 1, the structure of the holonomy is rather well understood in many situations because the structure of an orientation preserving diffeomorphism of $\mathbb{R}^1$ is known: there is a closed set of fixed points, and a countable collection of intervals of monotonicity. Consequently, any orientation preserving diffeomorphism exept the identity is of infinite order in the group of differmorphisms. Also, an isolated fixed point is bounded by two intervals of monotonicity. If a fixed point is not isolated, then the diffeomorphism has infinite contact with the identity at the fixed point, so that if the diffeomorphism is real analytic, it must be the identity. An orientation reversing diffeomorphism has a unique fixed point, and may most easily be studied by looking at its composition with the reflection about that fixed point.

In particular, interesting results can be obtained about codimension 1 foliations for which all the holonomy groups are trivial. Such a foliation necessarily is transversally orientable. The following definition is useful for the study of its holonomy pseudogroup.

5.9 Definition. A leaf has a *locally infinite holonomy pseudogroup* if given any transversal through a point P of the leaf, and any neighborhood U of P in the transversal, there is an orientation preserving element of the holonomy pseudogroup which is not a restriction of the identity and whose domain is contained in U.

This definition is due to Sacksteder and Schwartz [1965]. According to Example 5.16 below, a leaf may have a locally infinite holonomy pseudogroup, but a trivial holonomy group. However, the following Proposition makes the concept useful in codimension 1.

5.10 Proposition. *In codimension* 1, *if a leaf L of a C^1 foliation has a locally infinite holonomy pseudogroup, then any open set which meets L also meets a leaf with infinite holonomy group.*

Proof. Either the given element or its inverse must move some point Q toward P. The endpoint of the interval of monotonicity that contains Q lies on a leaf with infinite holonomy. □

Sacksteder and Schwartz give a number of conditions which imply the existence of a leaf with a locally infinite holonomy pseudogroup. Their proofs all depend on the following lemma. To state the lemma, it is convenient to introduce the notion of projector.

5.11 Definition. Given a C^1 codimension 1 foliation on a Riemannian manifold M, introduce a C^1 line element field transverse to the leaves and the family of curves obtained by integrating this field. Then a *projector* P is a continuous map from $[a,b]\times[0,S]$ to M such that for each (t,s)

(i) $P(t,s)$ lies in the intersection of the leaf through $P(0,s)$ and the transverse curve through $P(t,0)$.

(ii) $s\mapsto P(0,s)$ is an isometry.

5.12 Lemma. *Let P in M be such that the closure of the leaf through P is compact. Let*

$$P_i\colon [0,b_i]\times[0,S_i]\to M$$

be a sequence of projectors such that $P_i(0,0)=P$ for all i, $P_i(0\times(0,S_i))$ does not meet *the leaf through P, S_i tends to 0, and the length of $P_i(b_i \times [0, S_i])$ tends to a positive limit. Then the leaf through P has a locally infinite holonomy pseudogroup.*

Proof. See Sacksteder and Schwartz [1965]. □

This lemma has a number of useful consequences. One which will be required in § IV.5 follows.

5.13 Proposition. *Let L be a proper leaf with compact closure. Suppose further that there exists a leaf L′ such that L is contained in the intersection of the closures of the sets $L' \backslash K$, where K ranges over the compact subsets of L′. Then L has a locally infinite holonomy pseudogroup.*

Proof. See Sacksteder and Schwartz [1965]. □

This section will be completed by introducing some examples of codimension 1 foliations of compact 3-manifolds which show that very complicated leaves and holonomy can arise even in this apparently simple case.

5.14 Example. Let M_g be a compacted oriented 2-manifold of genus g, and choose on M_g two systems $\{\alpha_i\}$ and $\{\beta_i\}$ of g disjoint oriented embedded circles such that α_i crosses β_i once positively and meets no other curves of either system, and the β_i are disjoint. Let f_i be an orientation preserving diffeomorphism of S^1, and $p: M_g \times S^1 \to M_g$ be the projection. Cut $M_g \times S^1$ open along $p^{-1}(\beta_i)$, apply f_i to each fiber and paste together again. The resulting manifold is still $M_g \times S^1$, since each f_i is isotopic to the identity. It is foliated transversally to the fibers, with leaves obtained from sets of the form $M_g \times \{P\}$ by cutting and pasting. The holonomy along β_i is trivial, while the holonomy along α_i is obtained from f_i. Since the f_i are arbitrary, examples of foliations with complicated holonomy can be obtained in this way. (From the algebraic point of view, the f_i can be chosen arbitrarily because the α_i are free generators of a free subgroup of the fundamental group of M_g.)

5.15 Example. In Example 5.14 with $g = 2$, C^∞ diffeomorphisms f_1 and f_2 can be so chosen that there is a minimal set consisting of more than one leaf which is nowhere dense. For the details, see Sacksteder [1964].

5.16 Example. In Example 5.13 with $g = 2$, C^∞ diffeomorphisms f_1 and f_2 can be so chosen that there is a leaf with a locally infinite holonomy pseudogroup, but which has a trivial holonomy group. For the details, see Imanishi [1974].

6. Examples of Foliations

The object of this section is to present a number of examples of historical importance which still provide much of the motivation for the study of foliations. For the most part, they are foliations of low dimension or codimension, which can be visualized with varying degrees of ease.

The example of a nonzero vector field has already been mentioned in § 2. On a compact manifold, a vector field defines a flow, so that the vast theories of topological dynamics, differentiable dynamics, and measure preserving transformations are applicable. Many interesting properties of foliations have been discovered as generalizations of properties of flows.

The bundle with discrete structure group has already been discussed in §§ 2 and 5, where its usefulness as a source of computable examples is evident.

Any homogeneous space of a Lie group can be looked upon as a foliation, though it is not usually so viewed. When a Lie group G acts transitively on a manifold M, the set of elements which leave some arbitrarily chosen point P fixed forms a closed subgroup H. Hence, M can be identified with G/H, and those properties of G which are invariant under H can be carried down to M. If H is compact, averaging over H produces many invariant objects, but otherwise there may be few. In the latter case, it may be more convenient to consider the foliation of G by H than the quotient. For foliation theory, the most interesting case is the further generalization in which H is not required to be a closed subgroup, but merely to be generated by a subalgebra of the Lie algebra of G. A number of the important examples below are obtained by this method.

In codimension 1, new foliations can be constructed out of given ones by turning a transversal submanifold into a leaf and making the leaves which originally cut it spiral into it. The foliation induced on the transversal must be rather uniform for the construction to work. Fortunately, on a compact manifold, there are always transversal circles, and the boundary of a tube about one of them has the required properties. The precise statement of the construction is given by the following lemma.

6.1 Lemma. *Let M be an m-manifold foliated in codimension* 1, *and let N be a compact, oriented $(m-1)$ dimensional submanifold with trivial normal bundle. Suppose that in the neighborhood of N, the foliation is defined by a closed* 1-*form whose restriction to N is nonzero. Then there exists a foliation of M which agrees with the given foliation outside a neighborhood of N, is transversally oriented if the given foliation is, and has N as a leaf with holonomy of infinite order along some curve.*

Proof. The idea is that a foliation defined by a closed form admits a 1-parameter family of automorphisms defined by a nonzero vector field. The foliation can be bent in the direction of this flow until it becomes tangent to N. In fact, the details are formulated most neatly by not using this flow explicitly, but instead using the preliminary construction which will now be described. It will be assumed that all data are C^∞, and a C^∞ foliation will be constructed.

The tangent bundle to M at points of N is spanned by the tangent bundle to N and the tangent bundle to the foliation. Since the normal bundle of N is trivial, there is an embedding f of $N \times (-\varepsilon, \varepsilon)$ onto a neighborhood U of N contained in the set where ω is defined, such that $f(x, 0) = x$ for $x \in N$ and $f(\{x\} \times (-\varepsilon, \varepsilon))$ is contained in a leaf of the foliation. Let $\eta\colon \mathbb{R} \to \mathbb{R}$ be a C^∞ function such that

$$\eta(t) = \begin{cases} 1 & t \geqq \frac{1}{2}\varepsilon \\ 0 & t \leqq 0 \end{cases}$$
$$\eta'(t) > 0 \qquad 0 < t < \tfrac{1}{2}\varepsilon.$$

Then, if t denotes the parameter transverse to N in the product structure $N x(-\varepsilon, \varepsilon)$, the 1-form

$$\theta = \eta(t)\,\omega + (1 - \eta(t))\,dt$$

is nonzero and integrable. The C^∞ foliation it defines on U agrees with the given one for $t \geqq \frac{1}{2}\varepsilon$, and has as leaves the sets $N \times \{t\}$ for $t \leqq 0$. The holonomy map along any curve in N is the identity for $t \leqq 0$, so has the same derivatives as the identity for $t = 0$. On the other hand, there are curves along which the holonomy is not trivial. To see this, consider a vector field X on N such that $\omega(X) = 1$ everywhere. Any integral curve of X has a point of accumulation, hence can be modified slightly to produce a closed curve $\gamma\colon S^1 \to N$ such that the value of ω on each tangent vector to γ is positive. By using γ and the embedding f, one constructs an embedding

$$F\colon S^1 \times (-\varepsilon, \varepsilon) \to U.$$

Let $x \in \mathbb{R}/\mathbb{Z}$ be a parameter on S^1. Then $F^*\omega = h(x)\,dx$ for some positive function h on $\mathbb{R}/\mathbb{Z}$, so

$$F^*\theta = \eta(t)\,h(x)\,dx + (1 - \eta(t))\,dt.$$

$t = 0$ is an integral curve of $F^*\theta$, while the integral curves for $t > 0$ are obtained as solutions of the differential equation

$$\frac{dt}{dx} = -\frac{\eta(t)\,h(x)}{1 - \eta(t)} \tag{6.2}$$

Since the right hand side is strictly negative for $0 < t < \frac{1}{2}\varepsilon$ and all x, any solution passing through $(0, t_0)$ for $0 < t_0 < \frac{1}{2}\varepsilon$ exists for all positive values of x, and moves toward the curve $t = 0$ as x goes to infinity. Thus, the holonomy on the positive side of γ has no fixed points, each point approaches 0 as the map is iterated, and the map has infinite order.

There are a number of simple variants of the preceding construction. To bend the leaves in the opposite direction, change the sign of one term in θ. To construct a foliation which is trivial for positive t but nontrivial for negative t, replace t by $-t$. Because the holonomy along N is infinistesimally trivial, any two of these examples which are nontrivial on opposite sides of N can be combined to provide a C^∞ foliation of U which agrees with the given one near the boundary of U. (Alternatively, the manifold M can be cut open along N and the resulting boundary components glued to some other foliated manifold having a boundary leaf diffeomorphic to N and infinitesimally trivial holonomy.)

In order to obtain a foliation which is transversally orientable, it is necessary to be careful about the relation between the directions of bending. A transverse orientation on U is defined by a vector field X such that $\omega(X) = 1$. There is a vector field Y on U such that $Y = X$ for $t \geqq \frac{1}{2}\varepsilon$ and $\theta(Y) = 1$. Then for $t = 0$,

$dt(Y) = 1$. The construction using

$$\eta(-t)\omega + (1 - \eta(-t))\,dt$$

also gives $dt(Y) = 1$, so combining these two gives a transversally oriented foliation on U, and also on M if the given foliation is transversally oriented consistently with the transverse orientation on U. (As one proceeds along a curve γ in N with infinite holonomy, the leaves approach on one side and depart on the other.) This completes the proof of Lemma 6.1. □

In order to construct a foliation which is transversally orientable and has holonomy which is not infinitesimally trivial, it is possible to insert two compact leaves, in such a way that on each copy of γ, the holonomy is either contracting on both sides or expanding on both sides. Then the construction can be modified to obtain a foliation which is pictorially similar, but the holonomy map has first derivative different from 1. An example with one compact leaf and nontrivial linear holonomy is given by

$$\phi = \zeta(t)\omega + (1 - \zeta(t))\,dt$$

where ζ satisfies

$$\zeta(t) = \begin{cases} 1 & t \geqq \frac{1}{2}\varepsilon \\ 0 & t = 0 \\ -1 & t \leqq -\frac{1}{2}\varepsilon. \end{cases}$$

$$\zeta'(t) > 0 \qquad -\tfrac{1}{2}\varepsilon < t < \tfrac{1}{2}\varepsilon.$$

Then near $t = 0$

$$d\phi = -\zeta'(t)(1 - \zeta(t))^{-1}\omega \wedge \phi$$

so that the logarithm of the linear holonomy is given by the nonzero quantity

$$-\zeta'(0)\int_\gamma \omega.$$

Since any function f on N attains a maximum and $df = 0$ at a maximum, ω is not exact. Hence, de Rham's theorem insures the existence of an immersed circle γ such that the integral of ω along this circle is nonzero. However, de Rham's theorem does not imply that γ can be so chosen that the restriction of ω to γ is nowhere zero, a property that is required to obtain the desired configuration of leaves.

Since ω is closed, any vector field X such that $\omega(X) = 1$ satisfies $\mathscr{L}_X \omega = 0$, hence is an infinitesimal automorphism of the foliation defined by ω. Even more is true – the modified foliation also satisfies $\mathscr{L}_X \theta = 0$. As has already been remarked, the construction given in the lemma may be viewed as bending back the leaves of the given foliation in the direction specified by X. In terms of coordinates, these same facts are expressed by the observation that in the differential equation (6.2) the variables are separable.

6.3 Example. The two dimensional torus. The only oriented 2-manfiold which admits a 1-dimensional foliation is the torus $T^2 = S^1 \times S^1$. The suspension of any orientation preserving diffeomorphism of S^1 (as described in Example 5.2) is a foliation of T^2. Moreover, each circle $S^1 \times \{t\}$ is transverse to the suspension foliation, which is definable by a closed form in the neighborhood of any such transversal. In general, the compact leaves of the suspension correspond to the periodic points of the diffeomorphism, but in the case of diffeomorphisms of S^1, all periodic points have the same period. Any finite collection of disjoint transverse circles may be taken to be N in the lemma, and the bending may be performed in either direction at each circle. Thus, many foliations with isolated compact leaves but which are not suspensions may be produced. In particular, every homotopy class of line fields on T^2 contains such a foliation (Reinhart [1959c]). On the other hand, every foliation without compact leaves is the suspension of a diffeomorphism without periodic points (Siegel [1945]). If the diffeomorphism is of class C^k, $k = 0, 1$, there may be a nowhere dense minimal set, but if $k \geqq 2$, every leaf is dense (Denjoy [1932]). If every leaf is dense, the foliation is topologically equivalent to a foliation defined by a closed form (van Kampen [1935]). In a suitable coordinate system, such a foliation lifts to a family of parallel straight lines on the universal covering space $\mathbb{R}^2$, hence is called linear. (It is a homogeneous foliation, in the sense of Example 6.6.) In terms of diffeomorphisms, the theorem states that every C^2 diffeomorphism of the circle without periodic points is topologically equivalent to a rotation. On the other hand, a diffeomorphism of class C^k is usually not C^k conjugate to a rotation. This and related results have been obtained by Herman [1976].

Historically, Denjoy's result was the first in this area, and his methods are still being generalized. His idea is as follows: Suppose a diffeomorphism has no periodic points, and no orbit is dense. Then the complement of the minimal set is a countable union of intervals. The images of any such interval under the iterates of f are disjoint, hence have finite total length. On the other hand, the length of the image of an interval depends on the first derivative of f, which cannot vary too rapidly if the second derivative is continuous. This produces infinite total length, a contradiction. The same idea applied to the holonomy pseudogroup on a transverse interval enabled Schwartz [1963] to prove that the only minimal sets for a C^2 vector field on any compact 2-manifold are a point, a circle, and a torus. The statement that a foliation of codimension 1 is defined by a closed form is similar in content to the statement that the holonomy pseudogroup leaves invariant an absolutely continuous measure. Using this idea, Sacksteder has generalized the theorem of existence of such a form to C^2 foliations of codimension 1 and arbitrary dimension whose leaves have trivial holonomy. The precise statement and proof of his result will be given in Chapter IV, after the necessary techniques have been developed. This theorem has important applications to the study of the structure of such foliations.

6.4 Example. Unsmoothable foliations. Consider again a C^1 foliation of T^2 with no compact leaves and a nowhere dense minimal set. This foliation cannot be topologically conjugate to a C^2 foliation because the topological conjugacy would take the minimal set to a nowhere dense minimal set, contradicting Denjoy's theorem. Similar examples can be constructed on open 2-manifolds of higher genus, as follows: Remove some disks from the complement of the minimal set on some foliated tori with nowhere dense minimal sets. Glue in a number of foliated punctured annuli such that the foliations match on the boundary. Since the nowhere dense minimal sets are unaffected, Schwartz's theorem implies unsmoothability. For higher codimension, the situation is more complicated. For any finite k, Harrison [1975, 1979] has constructed examples of C^k diffeomorphisms of any compact manifold of dimension at least 2 which are not topologically conjugate to a C^{k+1} diffeomorphism. Suspending these diffeomorphisms gives examples of unsmoothable foliations of any codimension greater than 1.

6.5 Example. Three dimensional manifolds. The first example of a foliation of codimension 1 of a 3-manifold was given by Reeb [1952]. It can be constructed as follows: Let $S^1 \times \mathbb{R}^2$ be foliated with leaves $\{x\} \times \mathbb{R}^2$, and take N in Lemma 6.1 to be $S^1 \times C$ for some embedded circle C in $\mathbb{R}^2$. C together with its interior is a disc D^2, and $S^1 \times D^2$ inherits a foliation of codimension 1 having the boundary as the only compact leaf. A solid torus $S^1 \times D^2$ with this foliation is known as a Reeb component. The importance of such components arises from the fact that one can be constructed any time a circle transverse to a foliation is given. Since the holonomy has infinite contact with the identity, two such Reeb components may be joined by a diffeomorphism of the boundary to form a foliated compact 3-manifold; in particular, a foliation of the sphere S^3 may be obtained in this way. Any compact oriented 3-manifold without boundary admits a foliation of codimension 1 (Zieschang [1965], Lickorish [1965]), and there is even a foliation such that the tangent plane field belongs to any preassigned homotopy class (Wood [1969]). All of the constructions mentioned so far give foliations with Reeb components, and indeed, if the fundamental group of the manifold is finite, every foliation has a subset homeomorphic to a Reeb component (Novikov [1965], see also Haefliger [1967/68]). On the other hand, the 3-dimensional torus $T^3 = S^1 \times S^1 \times S^1$ has linear foliations such that every leaf is dense, so that none are compact (see Example 6.6 for details).

6.6 Example. Homogeneous foliations. Let G be a Lie group and $\mathfrak{g}$ its Lie algebra of left invariant vector fields. A p-dimensional subspace $\mathfrak{h}$ of $\mathfrak{g}$ defines a p-dimensional foliation of G if and only if $\mathfrak{h}$ is a subalgebra, whether or not the corresponding subgroup H is closed. A left invariant form ω may be identified with an element of the dual space $\mathfrak{g}^*$, and satisfies

$$d\omega(X, Y) = -\,\omega([X, Y])$$

for every pair (X, Y) of left invariant vector fields. Hence, $\mathfrak{h}$ is a subalgebra if and only if its annihilating subspace $\mathfrak{h}^\perp$ satisfies the condition: $d\mathfrak{h}^\perp$ is contained in the ideal generated by $\mathfrak{h}^\perp$ in the exterior algebra generated by $\mathfrak{g}^*$. Since $\mathfrak{h}$ generates a left invariant foliation, all the leaves are diffeomorphic.

Consider first the case of an abelian Lie group G. Any such group is a quotient of some vector group $\mathbb{R}^m$ by a discrete subgroup, hence is a product of a vector group and a torus T^n, where T^n is the product of n circles, $n \leqq m$. Since all the brackets in $\mathfrak{g}$ vanish, every subspace is a subalgebra. Each leaf is a product of a euclidean space and a torus, and so is its closure. If $G = T^m$, then the closure of each leaf is a torus, possibly of lower dimension. Since $\mathfrak{g}^*$ consists entirely of closed forms, these foliations are defined by closed forms, and each leaf has trivial holonomy. For computational purposes, it is convenient to think of T^m as the quotient of $\mathbb{R}^m$ by $\mathbb{Z}^m$. If $\mathbb{R}^m$ has coordinates $(x^1, \dots, x^m)$, then $\mathfrak{g}$ is generated by $\{\partial/\partial x^1, \dots, \partial/\partial x^m\}$ and $\mathfrak{g}^*$ is generated by $\{dx^1, \dots, dx^m\}$. Note that on T^m, dx^i is not exact, even though the notation makes it appear to be.

In general, for a noncompact Lie group G, one constructs foliations on compact manifolds which are obtained as quotients of G by discrete subgroups D. Since the left operation of G on itself commutes with the right operation, a left invariant foliation on G induces a foliation on the right quotient G/D. Since D is not usually normal, G/D is not a group, and the leaves on the quotient are not in general diffeomorphic.

From the viewpoint of Lie groups, the sharpest contrast to an abelian group is a simple group. The isometry group of hyperbolic n-space H^n is one such group which has interesting foliations associated with it. Actually, the most interesting of these are foliations of the unit tangent bundle to a compact manifold of constant negative curvature, but to describe these it is best to being with a description of the above-mentioned group. Let $\mathbb{R}^{n+1}$ have coordinates $(x^0, \dots, x^n)$, and consider the quadratic form $(x, y) = -x^0 y^0 + x^1 y^1 + \dots + x^n y^n$. This is nondegenerate quadratic form, but indefinite. (The necessary algebraic facts about such forms may be found, for example, in Artin [1957] or Wolf [1967]). Let $H^n = \{x \mid (x,x) = -1, x^0 > 0\}$. H^n is diffeomorphic to $\mathbb{R}^n$ by the projection onto the plane $x^0 = 0$, and its tangent space at x is the orthogonal space to x with respect to the quadratic form. Hence, the induced metric is Riemannian. H^k is embedded in H^n as the subset satisfying $x^{k+1} = \dots = x^n = 0$. This subset is totally geodesic because dropping the last $(n-k)$ components of a curve can only shorten it; in particular, H^1 is a geodesic. The group of linear transformations which leaves this quadratic form unchanged is denoted by $O(1,n)$, and the subgroup which also preserves the sign of x^0 is contained in the isometry group of H^n. The following analysis of the Lie algebra of $O(1,n)$ shows, among other things, that the identity component of $O(1,n)$ is also the identity component of the isometry group of H^n.

Since $O(1,n)$ is given as a subgroup of the general linear group of $\mathbb{R}^{n+1}$, the elements of its Lie algebra can be represented by matrices and the corresponding 1-parameter subgroups by matrix exponentials. These matrices will be thought of as acting on column vectors. In accord with previous notation, the rows and columns will be indexed from 0 to n. Since the maximum dimension of the isometry group of an n-dimensional Riemannian manifold is $\frac{1}{2}n(n+1)$, it will suffice to find this many linearly independent matrices whose exponentials preserve the quadratic form. In fact, these matrices will all be of the form

$$\begin{pmatrix} 0 & A \\ A^t & B \end{pmatrix}$$

where A is $1 \times n$ and B is $n \times n$ skew symmetric. The desired basis consists of matrices

$$\xi_i \qquad 1 \leqq i \leqq n$$

$$\zeta_{ij} \qquad 1 \leqq i < j \leqq n$$

each of which has two nonzero entries. ξ_i is symmetric and has a 1 in the $(0,i)$

place, while ζ_{ij} is skew-symmetric and has a -1 in the (i,j) place. The commutators of these elements are given by

$$[\xi_i, \xi_j] = -\zeta_{ij}$$

$$[\xi_i, \zeta_{jk}] = \begin{cases} -\xi_k & \text{if } i=j \\ +\xi_j & \text{if } i=k \\ 0 & \text{otherwise} \end{cases}$$

$$[\zeta_{ij}, \zeta_{jk}] = -\zeta_{ik}$$

$$[\zeta_{ij}, \zeta_{kl}] = 0 \qquad \text{if } i, j, k, l \text{ are distinct.}$$

In these formulas, $\zeta_{ji} = -\zeta_{ij}$ if $j \geqq i$. The 1-parameter group generated by ξ_i takes $(1, 0, \ldots, 0)$ to the points in the (x^0, x^i) plane with coordinates $(\cosh t, \sinh t)$, so the group acts transitively on H^n. Note that $\{\zeta_{ij}\}$ generates a subgroup which is the group of euclidean rotations in $\mathbb{R}^{n+1}$ leaving the x^0-axis fixed. Hence, the full group acts transitively on tangent 2-planes, and therefore H^n has constant sectional curvature. Even more can be said: the identity component of $O(1,n)$ acts on the tangent n-frames in such a way that exactly one group element takes any given oriented orthonormal frame to any other given such frame. Hence, the identity component is diffeomorphic to the bundle of oriented orthonormal n-frames on H^n.

The group $O(1,n)$ has two foliations of dimension n, given by the subalgebras with bases

$$\left\{\xi_1, \frac{\sqrt{2}}{2}(-\xi_2 + \zeta_{12}), \ldots, \frac{\sqrt{2}}{2}(-\xi_n + \zeta_{1n})\right\}$$

$$\left\{\xi_1, \frac{\sqrt{2}}{2}(-\xi_2 - \zeta_{12}), \ldots, \frac{\sqrt{2}}{2}(-\xi_n - \zeta_{1n})\right\}$$

The bundle $T_1(H^n)$ of unit tangent vectors to H^n is diffeomorphic to the quotient of the isometry group by any subgroup leaving a point and a unit tangent vector at that point fixed. The subgroup corresponding to the subalgebra $\{\zeta_{ij} \mid 2 \leq i < j\}$ is such an isotropy subgroup, which also leaves the above two foliations invariant under its adjoint action. Consequently, these two foliations pass to the right quotient, giving rise to foliations F^+ and F^- of dimension n and codimension $n-1$. Since the intersection has dimension 1, these two foliations are transverse to each other. The flow generated by the vector field ξ_1 commutes with the isotropy subgroup, so ξ_1 passes to a vector field on $T_1(H^n)$ lying in the intersection of the two foliations. Since $O(1,n)$ is defined as a set of transformations of H^n, the identity

$$g_1(g_2(x)) = (g_1 g_2)(x)$$

holds for all $g_1, g_2 \in O(1,n)$ and $x \in H^n$. Thus the cosets of any 1-parameter subgroup have isometric orbits in H^n. In particular, since H^1 is an orbit of ξ_1 and H^1 is a geodesic, all the orbits of ξ_1 are geodesics. Furthermore, since each 1-parameter subgroup acts by isometries, each has orbits parameterized proportionally to arc length. Each leaf of F^+ is foliated by curves which are projections of cosets of ξ_1 and project into geodesics on H^n. For a geometric understanding of F^+, the crucial problem is to characterize the configuration in H^n obtained by projecting all the curves lying on a single leaf of F^+. Let ξ and η_i denote the vector fields ξ_1 and $(\sqrt{2}/2)(-\xi_i + \zeta_{1i})$ on $O(1,n)$ respectively. Then for $2 \leqq j \leqq k$

$$[\xi, \eta_i] = \eta_i \qquad [\xi, \zeta_{jk}] = 0$$

$$[\eta_i, \zeta_{jk}] = \begin{cases} -\eta_k & i = j \\ \eta_j & i = k \\ 0 & \text{otherwise.} \end{cases}$$

Let η denote any vector field of the form $\sum r^i \eta_i$ for $r^i \in \mathbb{R}$. Since $[\xi, \eta] = \eta$, ξ and η generate a family of 2-dimensional submanifolds in $O(1,n)$, and the one passing through the identity is a subgroup whose orbits foliate H^n and lie in the projection of a single leaf of F^+. Moreover, it can be verified by examining the orbit through $(1,0,\dots,0)$ that the ξ orbits in the (ξ,η) surface diverge exponentially in the positive direction and converge exponentially in the negative direction. (The equation $\mathscr{L}_\xi \eta = \eta$ has the same implications as the equation $dx/dt = x$.) The final picture is that each leaf of F^+ projects diffeomorphically onto H^n, in such a way that the family of ξ curves projects to the family of geodesics asymptotic in the negative direction to any one of its members. For F^-, the result is similar, except that the geodesics are asymptotic in the positive direction. Since each leaf of F^+ is a closed subset diffeomorphic to $\mathbb{R}^n$, each is a minimal set and has trivial holonomy.

The constant curvature of H^n must be negative, because of the existence of these families of diverging geodesics. For each n, there exist compact manifolds M of dimension n and constant negative curvature. By multiplying the metric teasor by a constant if necessary, the universal covering space of such a manifold may be taken to be H^n. The fundamental group $\pi_1(M)$ may therefore be considered as a subgroup of the isometry group of H^n which has only the identity in common with the subgroup generated by $\{\zeta_{ij}\}$. The quotient of $T_1(H^n)$ by the left action of $\pi_1(M)$ is the tangent sphere bundle $T_1(M)$, and the foliations are left invariant, so they induce foliations of $T_1(M)$. The minimal sets and holonomy of these foliations are extremely complicated, as the case $n = 2$ shows. In this case, M and the compact leaves of either foliation are surfaces of genus g at least 2, so the fundamental groups are given by $2g$ generators and one relation. For the noncompact leaves, the fundamental groups are free. Every free homotopy class on M contains a closed geodesic, which is covered by a closed integral curve of ξ_1 that cannot bound a disc on its leaf because ξ_1 is nonzero. Hence, there exist nonsimply connected leaves, and indeed, there are leaves with nonabelian fundamental group. This fact allows the holonomy to become very complicated, even though the codimension is 1.

A flow corresponding to ξ_1 can be defined on the tangent sphere bundle of any Riemannian manifold. Indeed, there is a unique geodesic through any point of a Riemannian manifold with any given initial unit vector, so the geodescis lift to a family of curves forming a 1-dimensional foliation of the tangent sphere bundle. This can be made into a flow by using the natural parametrization of the geodesics. It is a classical problem to understand the nature of this flow, a problem which is very difficult even for compact surfaces of constant negative curvature.

Anosov [1963] used the expanding and contracting foliations to obtain information on the nature of this flow. The foliations also exist in the case of variable negative curvature, but are not smooth in the transverse direction. Thus, the generalization to this and other cases encounters technical difficulties but, in spite of them, has been largely carried out and been useful.

There are a number of problems in which it is important to study the set of points that can be reached from a given point by curves tangent to a given plane field. The plane field may be defined by forms or by vector fields, and the dimension may vary from point to point. In the case of a foliation, the set of points reachable from a given point makes up the leaf through that point. This is in some sense the most restricted case, and much larger reachable sets occur commonly. Two specific cases will be discussed here: thermodynamics and control theory.

6.7 Example. Thermodynamics. Carathéodory [1909] gives a treatment of the foundations of thermodynamics in which the key hypothesis is the existence near any initial state of states which cannot be reached from that initial state. In this treatment, the set of equilibrium states of an isolated system is assumed to be parametrized by some $\mathbb{R}^m$ whose coordinates are interpreted as volumes, pressures, and densities. Changes of state take place along piecewise differentiable curves γ in $\mathbb{R}^m$, and the external work supplied during such a change is the integral along the curve of a given 1-form ω. There are two axioms. The first asserts the existence of a function E, the internal energy, which can be changed only by external work supplied or done. The second asserts that in any neighborhood of any given state, there are states which cannot be reached from it. A curve along which $dE + \omega = 0$ is called a quasi-static change, because physically it is viewed as proceeding so slowly that there is no loss of available energy through friction. The following proposition implies that there exist functions S and T such that

$$dE + \omega = T\,dS. \tag{6.8}$$

6.9 Proposition. *Let the C^1 form in $\mathbb{R}^m$*

$$dx^1 + f_2(x)\,dx^2 + \ldots + f_m(x)\,dx^m = \phi$$

be such that in every neighborhood of every point, there exist points which cannot be reached by integral curves of this form. Then there exist functions T and S such that $\phi = T\,dS$.

Proof. See Carathéodory [1909, pp. 369–370]. One first constructs the foliation of $\mathbb{R}^m$ by level surfaces of S, then from it obtains S and T. □

This beautiful result has still not used the full strength of the second axiom. Consider now a change of state $\gamma(t)$ which need not be quasi-static. The value of dS on a tangent vector to $\gamma(t)$ can be zero, but the sign of its nonzero values must be independent of γ, since otherwise every point could be reached. In the physical interpretation, T is absolute temperature and S is entropy, so with the proper choice of signs, this last result says that entropy never decreases, and that it remains constant under quasi-static changes. The latter are therefore the only reversible changes.

In order to make a universal determination of S and T, one takes two isolated systems and puts them into contact in such a way that one relation among the variables, is introduced. By applying Proposition 6.8 to each system separately and to the combined system it is possible to determine T up to a multiplicative constant, which in turn determines S up to an additive constant. Besides Carathéodory's paper, the book of Born [1949] gives a good presentation of these ideas. The statement of the second axiom which is given here is Born's version. Carathéodory is careful to observe that from a physical point of view, one should speak of approximating a given state rather than of reaching it, since no measurement can be precise enough to make certain that the state is reached.

6.10 Example. Control theory. The problem is control theory is to determine whether some physical system in a given state can be brought to some other state, and if go by what paths and in how much time. For example, in the theromdynamical problem just considered, it is possible to get from one state to another and back again if and only if they lie on the same leaf of the foliation, which occurs precisely if they have the same entropy. Any path which stays on the leaf is possible. If the requirement of reversibility is dropped, then the admissible paths are those along which entropy is nondecreasing, and a final state is reachable if and only if it has entropy not less than the initial state. In general, one cannot expect that the problem will be representable in terms of a single Pfaffian equation, or even that the number of equations needed will be the same from point to point. The following paragraphs contain the mathematical description of a fairly general type of system of interest in control theory. The main results summarized here are due to Sussmann [1973] and Stefan [1974].

Let M be a manifold on which there is given a number of vector fields, each defined on an open subset. These vector fields generate a pseudogroup of diffeomorphisms of M, and it is required to determine the orbit of each point under this pseudogroup. Any point on a given orbit can be reached from any other by a continuous, piecewise smooth curve such that each smooth piece is an integral curve of one of the given vector fields. Since the orbit of a vector field is 1-dimensional, $(n-1)$ corners are required to reach every point of an n-dimensional orbit. Thus, a coordinate system is obtained locally by using the time coordinates of the $(n-1)$ break points and of the final point. To show that each orbit is actually a smooth submanifold, it is necessary to study its tangent space. This tangent space E_x at any point x contains the linear hull D_x of the set of vectors at that point belonging to the given vector fields. In addition, E_x contains the images under the pseudogroup of D_y for each point y on the orbit of x, and is in fact the linear hull of all these images. Thus, the dimension of E_y is constant on each orbit, though not necessarily on M. The collection of all the orbits constitutes a foliation with singularities, in the sense that each point x on an orbit of dimension p belongs to an open set U which is diffeomorphic to $\mathbb{R}^p \times \mathbb{R}^{m-p}$, such that the intersection of each orbit with U is the union of sets of the form $\mathbb{R}^p \times \mathbb{R}^j$ for some $j \geqq 0$.

In the development of this subject, there was an important step between Carathéodory's paper of 1909 and the recent papers of Sussmann and of Stefan. This was Chow's paper of 1940 on the integrability of plane fields of constant codimension, not necessarily 1. The methods of Chow's paper have been incorporated in § 4. They foreshadow the recent work in that the translates of each given vector field by the flows defined by the others are explicitly considered.

Chapter II. Prolongations, Connections, and Characteristic Classes

One aim of this chapter is to present some standard facts about connections and their relation to characteristic classes, and then show how these ideas extend to give new kinds of characteristic classes associated to a foliation. The presentation of the material will not be at all standard, however. Indeed, the same characteristic classes associated to a foliation of codimension q can be obtained from the algebra of formal power series vector fields at the origin in $\mathbb{R}^q$, without any reference to foliations at all. This result is consistent with the point of view that foliations are the natural quotient objects in differential geometry, but it also suggests an exposition in which higher order derivatives are incorporated from the beginning, in order to make the relation appear as clear and natural as possible. Thus, the first section is a discussion of the Lie groups of invertible polynomial mappings. The composition of two mappings of degree k is made into a mapping of degree k by discarding higher order terms, a procedure consistent with the application to Taylor series of differentiable mappings discussed in the second section. The higher order frame bundles on a manifold are principal bundles whose groups are truncated polynomial groups. In addition, each carries a Lie algebra valued 1-form, the tautological form, which contains precisely the information that the derivatives of various orders are not independent when more than one point is involved. Any G-structure has associated to it reductions of the higher order frame bundles to subgroups of the truncated polynomial groups. In this way, each subgroup of the general linear group has associated to it extensions which are polynomial groups. In the third section, these extensions are discussed. In the fourth section, the properties of connections in a principal bundle are summarized, especially the relation between connections and the characteristic classes of the bundle. In the case of $J^k M$, the bundle also carries the tautological form θ, and the properties of θ and the connection form ω are compared and contrasted. The exterior covariant derivative of θ is the torsion form Θ, and connections with $\Theta = 0$ are particularly interesting. Finally, in the fifth section, various connections associated to a foliation are introduced, and the characteristic classes of foliations are introduced from the connection point of view. (The relation to formal power series vector fields is discussed in a broader context in Chapter III.)

Since the truncated polynomial groups are neither semi-simple nor solvable, very little of existing Lie group theory applies to them. Since these are very important groups, the author hopes that this situation will not be permitted to continue.

1. Truncated Polynomial Groups and Algebras

If a real valued function of a real variable has k continuous derivatives at a point a, then in the neighborhood of a it may be written

$$f(a + x) = f(a) + f'(a)x + \ldots + \frac{1}{k!}f^{(k)}(a)x^k + R_k(x, a)$$

where $x^{-k}R_k(x,a)$ approaches 0 as x approaches 0. The sum, product, and composition of two such functions are given by similar expressions, where the right hand side is a polynomial of degree k plus a remainder satisfying the same growth condition. The polynomial part is found by applying the usual rules for sum, product, or composition of polynomials, then discarding the terms of degree greater than k. This operation of discarding all terms of degree greater than some given integer is called truncation. The study of truncated polynomial operations is important for the understanding not only of functions having a finite order of differentiability, but also of C^∞ functions, since the ring of formal power series is on the one hand the inverse limit of truncated polynomial rings, and on the other hand the ring of Taylor series of C^∞ functions. Analogous remarks hold for mappings from $\mathbb{R}^m$ to $\mathbb{R}^m$, provided derivative is interpreted in the sense of total derivative. In this section, the elementary properties of truncated polynomial groups and algebras will be reviewed. In particular, the Lie algebras of certain truncated polynomial groups are isomorphic to the algebras of derivations of formal power series rings, and these derivations can be written in a way which is very convenient for calculation.

Let $\mathbb{R}[x] = \mathbb{R}[x^1, \ldots, x^m]$ be the ring of polynomials in m variables with real coefficients. Under the operation of composition (formal or as real functions), $\mathbb{R}[x]^m$ is a monoid which operates on $\mathbb{R}[x]$ on the right by the usual formulas. This monoid has an identity, but in general an element does not have an inverse, even if its linear part is nonsingular as a linear mapping. In the latter case, there is an analytic mapping which is its inverse on some open set, but this open set cannot contain any points where the Jacobian determinant of the polynomial mapping vanishes. Since a polynomial is defined everywhere, the inverse cannot be a polynomial. Another difficulty with the monoid $\mathbb{R}[x]^m$ for present purposes is that the composition of C^k mappings is a C^k mapping, while the composition of polynomials of degree k is a polynomial of degree k^2. Both of these difficulties are eliminated by truncation, which is justified by the following lemmas.

1.1 Lemma. *If $f \in \mathbb{R}[x^1, \ldots, x^m]^n$ and $g \in \mathbb{R}[x^1, \ldots, x^n]^p$, then $g \circ f$ is defined and is an element of $\mathbb{R}[x^1, \ldots, x^m]^p$. If in addition f has all constant terms 0, then the terms of degree at most k of $g \circ f$ depend only on the terms of degree at most k of g and of f.*

Proof. To define $g \circ f$, replace each variable x^i in g by the i-th element of f. The result is a p-tuple of polynomials in $x^1, \ldots, x^m$. The last statement follows from the fact that for any polynomial p with constant term 0, p^{k+1} has no term of degree less than $k + 1$.

1.2 Lemma. *Let $f \in \mathbb{R}[x^1, \ldots, x^m]^m$ be such that each component has constant term 0 and degree at most k. Suppose further that the linear terms define a nonsingular mapping. Then there is a $g \in \mathbb{R}[x^1, \ldots, x^m]^m$ satisfying the same conditions such that $g \circ f$ and $f \circ g$ each differ from the identity by terms of degree at least $k + 1$.*

Proof. According to Bourbaki [Alg. IV, p. 64], given any polynomial mapping f of degree k with constant terms 0 and nonsingular linear terms, there is an m-tuple g of formal power series with constant terms 0 and nonsingular linear terms such that $(f \circ g)(x)$ is the m tuple $(x^1, \ldots, x^m)$, that is, the composition is the identity. Let g^k denote the polynomial mapping obtained by taking the terms of degree at most k in each component of g. By Lemma 1.1, $f \circ g^k$ differs from the identity by terms of degree at least $k + 1$. Hence, $g^k \circ f \circ g^k \circ f$ differs from $g^k \circ f$ by terms of degree at least $k + 1$. Let h be a formal power series such that $g^k \circ f \circ h$ is the identity. Then, for the terms of degree at most k, the following equations hold:

$$g^k \circ f \circ g^k \circ f \circ h = g^k \circ f \circ h, \qquad g^k \circ f = 1.$$

Hence g^k is an inverse for f, up to degree k. □

1.3 Definition. $GP_m^k = GP_m^k(\mathbb{R})$ is the group whose underlying set is the subset of $\mathbb{R}[x]^m$ consisting of elements with nonsingular linear terms, such that each component has 0 constant term and no term of degree greater than k. The operation assigns to the pair (f, g) the terms of degree at most k of $f \circ g$.

1.4 Definition. Π_k^l is the mapping from GP_m^l to GP_m^k, defined for $l \geqq k$ by taking each polynomial to the polynomial consisting of its terms of degree at most k. Π_k^l is called a *truncation operator*.

1.5 Definition. The *order* of a polynomial is the degree of its lowest degree nonzero term. $\mathbb{R}[x]_k$ is the quotient of $\mathbb{R}[x]$ by the ideal consisting of polynomials of order at least $k + 1$. The truncation homomorphism Π_k^l from $\mathbb{R}[x]_l$ to $\mathbb{R}[x]_k$, $l \geqq k$, is induced by the mapping which takes each polynomial of degree at most l to its terms of degree at most k.

1.6 Proposition. *GP_m^k is a Lie group under the given operation, and the Π_k^l are epimorphisms satisfying $\Pi_k^l \circ \Pi_l^m = \Pi_k^m$. There is an analytic action of GP_m^k on the right of $\mathbb{R}[x]_k$ which commutes with Π_k^l.*

The identity is the only element of GP_m^k that leaves every element of $\mathbb{R}[x]_k$ fixed. $GP^1{}_m$ is isomorphic to GL_m. The dimension of GP_m^k is

$$m \sum_{r=1}^{k} \binom{m + r - 1}{m - 1}.$$

Proof. The coefficients of the polynomials are analytic coordinates for the group, from which fact the rest of the statements follow. In particular, the formula for the dimension of GP_m^k follows from the fact that the number of distinct monomials of degree r in m commuting indeterminants is given by the binomial coefficient

$$\binom{m + r - 1}{m - 1}. \quad \square$$

1.7 Remark. A full analysis of the orbits of the action of GP_m^k on $\mathbb{R}[x]_k$ is quite difficult, and is not needed in this work. It is needed in the theory of singularities of smooth mappings, which owes many of its difficulties to the necessity of this analysis.

The topological group GP_m^∞ obtained as the inverse limit of the groups GP_m^k acts on the vector space $\mathbb{R}[\![x]\!]$ which is the inverse limit of the vector spaces $\mathbb{R}[x]_k$, and this action commutes with the natural projections. The elements of the limit spaces may be represented by infinite formal sums, and it is convenient to do so for calculation purposes.

1.8 Definition. N_m^k is the normal subgroup of GP_m^k obtained as the kernel of Π_1^k for $2 \leqq k \leqq \infty$.

Since GP_m^k is a Lie group if k is finite, a detailed knowledge of its structure is best obtained with the aid of its Lie algebra. Since GP_m^k acts on $\mathbb{R}[x]_k$, its Lie algebra acts as an algebra of derivations on the functions on $\mathbb{R}[x]_k$, hence also on the Taylor series at the origin of such functions. Since $\mathbb{R}[x]_k$ can be identified with the set of Taylor series up to the power k, one is led to study derivations of $\mathbb{R}[x]_k$. Though GP_m^∞ is infinite dimensional, it also has a Lie algebra in a natural sense, namely, the inverse limit of the Lie algebras of the groups GP_m^k. Thus, finally, one is led to study derivations of $\mathbb{R}[\![x]\!]$.

1.9 Definition. An *endomorphism* of the inverse system $\{\mathbb{R}[x]_k, \Pi_k^l\}$ is a familiy $\{D_k\}$ of homomorphisms such that the diagram

$$\begin{array}{ccc} \mathbb{R}[x]_k & \xrightarrow{D_k} & \mathbb{R}[x]_k \\ \Big\downarrow \Pi_l^k & & \Big\downarrow \Pi_l^k \\ \mathbb{R}[x]_l & \xrightarrow[D_l]{} & \mathbb{R}[x]_l \end{array}$$

is commutative. The limit D is a *derivation* if in addition it satisfies the product rule

$$D(fg) = (Df)g + f(Dg).$$

D is said to be *homogeneous of degree* p if for each r, each homogeneous polynomial of degree r is mapped into a homogeneous polynomial of degree $r + p$. D is *bounded* if each polynomial is mapped into a polynomial.

1.10 Proposition. *Let D be a bounded derivation of $\mathbb{R}[\![x]\!]$. Then D applied to any constant gives 0, and D is uniquely determined by the values $D(x^1), \ldots, D(x^m)$ which are arbitrary polynomials. D is homogeneous of degree p if and only if each $D(x^i)$ is a homogeneous polynomial of degree $p + 1$. D is the sum of finitely many homogeneous derivations, each of degree at least -1, and is so in exactly one way.*

Proof. Since $D(1) = D(1 \cdot 1) = D(1) + D(1)$, $D(1) = 0$ and hence $D(r) = rD(1) = 0$ for each real number r. Since D is an inverse limit of endomorphisms, it is an endomorphism, that is,

$$D\left(\sum_{r=0}^{\infty} a_{i_r \dots i_r} x^{i_1} \dots x^{i_r}\right) = \sum_{r=0}^{\infty} a_{i_1 \dots i_r} D(x^{i_1} \dots x^{i_r}),$$

where repeated indices are summed. Since

$$D(x^{i_1} \dots x^{i_r}) = \sum_{\alpha=1}^{r} x^{i_1} \dots x^{i_{\alpha-1}} D(x^{i_\alpha}) x^{i_{\alpha+1}} \dots x^{i_r},$$

D is determined as soon as all the polynomials $D(x^i)$ are known. Furthermore, given any collection of m polynomials, the above formulas define a unique mapping of $\mathbb{R}[\![x]\!]$ which satisfies all the necessary conditions to be a bounded derivation. D is the sum of a finite number of operations, each of which consists of deleting one of the variables x^i from a monomial and replacing it by the polynomial $D(x^i)$. In particular, if each $D(x^i)$ is homogeneous of degree $p+1$, this operation increases the degree of the monomial by p, while if some $D(x^i)$ is not homogeneous, then there is a monomial, namely x^i, whose image is not homogeneous. Since the degree of a homogeneous polynomial must take one of the values $\{0, 1, 2, \dots\}$ the degree of a homogeneous derivation must take one of the values $\{-1, 0, 1, \dots\}$. This completes the proof. □

1.11 Definition. ${}^p\mathfrak{a}_m$ is the space of homogeneous derivations of degree p of $\mathbb{R}[\![x^1, \dots, x^m]\!]$.

$$\mathfrak{a}_m^k = \prod_{p=-1}^{k-1} {}^p\mathfrak{a}_m \qquad \text{for } k \geqq 0.$$

$$\mathfrak{gp}_m^k = \prod_{p=0}^{k-1} {}^p\mathfrak{a}_m \qquad \text{for } k \geqq 1.$$

$$\mathfrak{n}_m^k = \prod_{p=1}^{k-1} {}^p\mathfrak{a}_m \qquad \text{for } k \geqq 2.$$

π_k^l: $\mathfrak{a}_m^l \to \mathfrak{a}_m^k$ is the natural projection, defined for $l \geqq k$, and $\mathfrak{a}_m^\infty$ is the inverse limit of the spaces $\mathfrak{a}_m^k$ under this system of projections. Also, $\mathfrak{gp}_m^\infty$ and $\mathfrak{n}_m^\infty$ are the inverse limits defined by the restrictions of $\prod_k^l$ to $\mathfrak{gp}_m^l$ and $\mathfrak{n}_m^l$.

$$f^i(x)\frac{\partial}{\partial x^i} = \sum_{i=1}^{m} f^i(x^1, \dots, x^m)\frac{\partial}{\partial x^i},$$

where $f^i(x)$ is a formal power series, is the element of $\mathfrak{a}_m^\infty$ whose projection D^k into the endomorphisms of $\mathbb{R}[x]_k$, $k \geqq 1$, is such that $D^k(x^i)$ is given by the terms of degree at most k in $f^i(x)$.

1.12 Proposition. $\mathfrak{a}_m^\infty$ *is homeomorphic to* $\prod_{p=-1}^{\infty} {}^p\mathfrak{a}_m$, *consists of all derivations of* $\mathbb{R}[\![x]\!]$, *and contains the bounded derivations as a dense subset. It is a Lie algebra under the bracket operation*

$$[D_1, D_2] = D_1 \circ D_2 - D_2 \circ D_1$$

for which the formula in coordinates is

$$[D_1, D_2] = \left(f^i \frac{\partial g^j}{\partial x^i} - g^i \frac{\partial f^j}{\partial x^i}\right) \frac{\partial}{\partial x^j}$$

where $D_1 = f^i \partial/\partial x^i$ and $D_2 = g^j \partial/\partial x^j$. This Lie algebra is graded, in the sense that if $D_1 \in {}^p\mathfrak{a}_m$ and $D^2 \in {}^q\mathfrak{a}_m$, then $[D_1, D_2] \in {}^{q+p}\mathfrak{a}_m$.

1.13 Corollary. $\mathfrak{gp}_m^\infty$ *and* $\mathfrak{n}_m^\infty$ *are subalgebras of* $\mathfrak{a}_m^\infty$. *The derivations of degree at least k form an ideal in either, and the quotient algebras give Lie algebra structures to* $\mathfrak{gp}_m^k$ *and* $\mathfrak{n}_m^k$. *The derivations of degree at least k do not form an ideal in* $\mathfrak{a}_m^\infty$. $\mathfrak{a}_m = {}^{-1}\mathfrak{a}_m \oplus {}^0\mathfrak{a}_m$ *is a subalgebra having* ${}^{-1}\mathfrak{a}_m$ *as an ideal.* $\mathfrak{gl}_m = {}^0\mathfrak{a}_m = \mathfrak{gp}_m^1$ *is a subalgebra, which is isomorphic to* $\mathfrak{a}_m/{}^{-1}\mathfrak{a}_m$.

Note that the operation of $\mathfrak{a}_m^\infty$ as a set of derivations of $\mathbb{R}[\![x]\!]$ induces an operation of $\mathfrak{gp}_m^k$ as a set of derivations of $\mathbb{R}[x]_k$. Since no element of GP_m^k except the identity acts trivially on $\mathbb{R}[x]_k$, $\mathfrak{gp}_m^k$ may be identified with the Lie algebra of GP_m^k. Furthermore, $\mathfrak{a}_m$ may be identified with the Lie algebra of the affine group A_m. As a consequence the Lie algebras of GP_m^∞, GL_m, N_m^k, and N_m^∞ may be identified respectively with $\mathfrak{gp}_m^\infty$, $\mathfrak{gl}_m$, $\mathfrak{n}_m^k$, and $\mathfrak{n}_m^\infty$.

Let $\mathfrak{n}_{m,l}^k$ be the kernel of Π_l^k, so that $\mathfrak{n}_{m,1}^k = \mathfrak{n}_m^k$. $\mathfrak{n}_{m,l}^k$ is nilpotent, and its lower central series can easily be computed by using bounded derivations of $\mathbb{R}[\![x]\!]$. The corresponding groups $N_{m,l}^k$ are connected, so their lower central series are determined by those of the corresponding Lie algebras.

An exact sequence

$$0 \to \mathfrak{n} \to \mathfrak{h} \to \mathfrak{g} \to 0$$

is also known as an extension over $\mathfrak{g}$ with kernel $\mathfrak{n}$. The properties of such extensions and their classification by means of cohomology theory are discussed by Cartan and Eilenberg [1956, Chapter 13 and 14]. If $\mathfrak{n}$ is abelian, $\mathfrak{h}$ is characterized up to the natural equivalence relation by the action of $\mathfrak{g}$ in $\mathfrak{n}$ and the cohomology group $H^2(\mathfrak{g}; \mathfrak{n})$, where $\mathfrak{n}$ is regarded as a $\mathfrak{g}$-module under this action. It would be useful to know more about the many extensions that relate the various subalgebras of $\mathfrak{a}_m^\infty$. The following proposition gives the principal properties in the simplest case.

1.14 Proposition. $\mathfrak{n}_{m,k-1}^k$ *for $k \geqq 2$ is an abelian Lie algebra, whose underlying vector space is mapped isomorphically onto $N_{m,k-1}^k$ by the exponential map. The action of* $\mathfrak{gp}_m^{k-1}$ *on* $\mathfrak{n}_{m,k-1}^k$ *depends only on* $\mathfrak{gl}_m$, *and under this action* $\mathfrak{n}_{m,k-1}^k$ *is isomorphic to $S^k(\mathbb{R}^m)^* \otimes \mathbb{R}^m$, where $(\mathbb{R}^m)^*$ is the dual space to $\mathbb{R}^m$ and S^k denotes the symmetrized tensor power. This* $\mathfrak{gl}_m$*-module is the direct sum of trace-free tensors and tensors of the form*

$$p(x^1, \ldots, x^m) \sum_{i=1}^m x^i \frac{\partial}{\partial x^i}$$

where $p(x)$ ranges over polynomials of degree $(k-1)$.

The element of H^2 $(\mathfrak{gp}_m^{k-1}; \mathfrak{n}_{m,k-1}^k)$ *defined by the extension* $\mathfrak{gp}_m^k$ *is 0 for $k = 2$, and for $m = 1$ and $k = 3$, but is nonzero otherwise.*

Proof. Much of the proof is elementary calculations with the derivations of $\mathbb{R}[\![x]\!]$. The direct sum decomposition of $\mathfrak{n}^k_{m,k-1}$ is due to Terng [1978], and much interesting related material can be found in her paper. The extensions mentioned in the last sentence have been studied by Reinhart [1982]. □

The Lie algebra statements imply corresponding statements about the groups. These have been used by Terng to classify the connected normal subgroups of GP^k_m.

Since $\mathfrak{gp}^\infty_m$ is the inverse limit of finite dimensional Lie algebras, it is natural to consider it as the Lie algebra of the corresponding inverse limit of groups. The study of $\mathfrak{a}^\infty_m$ is more complicated, since it is an inverse limit as a vector space but not as a Lie algebra, and there are no finite dimensional groups from which to construct an inverse limit. Indeed, since the set of derivations of degree at least k is not an ideal in $\mathfrak{a}^\infty_m$ for $k \geqq 1$, the quotient of $\mathfrak{a}^\infty_m$ by this subspace does not inherit the structure of a Lie algebra, and one therefore cannot look for the corresponding group. In order to state what is true about finite approximation to $\mathfrak{a}^\infty_m$, some further notation will be required.

Let $\mathfrak{a}^k_m$ denote the vector space which is described by

$$\mathfrak{a}^k_m = \mathfrak{gp}^k_m \oplus {}^{-1}\mathfrak{a}_m = \prod_{p=-1}^{k-1} {}^p\mathfrak{a}_m$$

1.15 Lemma. *There is an operation induced by the Lie bracket in $\mathfrak{a}^\infty_m$ which takes a pair of elements in $\mathfrak{a}^k_m$ to an element of $\mathfrak{a}^{k-1}_m$, is antisymmetric, and is such that the triple product*

$$(\mathfrak{a}^k_m \times \mathfrak{a}^k_m) \times \mathfrak{a}^k_m \to \mathfrak{a}^{k-2}_m$$

satisfies the Jacoby identity.

Proof. Since the degree of a derivation is at least -1, the bracket of an element of ${}^p\mathfrak{a}_m$ for $p \geqq k$ with an element of $\mathfrak{a}^k_m$ has degree at least $k-1$, hence is 0 in $\mathfrak{a}^{k-1}_m$. Using this fact, one can easily verify the statements made. □

For further study of $\mathfrak{a}^k_m$, an analog of the adjoint operation is useful. One considers conjugating a polynomial of degree $(k-1)$, which may have nonzero constant term, by a polynomial representing an element of GP^k_m. This gives rise to a bracket

$$\mathfrak{gp}^k_m \times \mathfrak{a}^{k-1}_m \to \mathfrak{a}^{k-1}_m$$

which is a restriction of the operation considered in Lemma 1.15. The details will be deferred to § 2, since additional machinery is required to state precisely and prove the result. On the other hand, the properties of the dual space $(\mathfrak{a}^k_m)^*$ are easily stated and proved, and this will be done immediately after the discussion of the dual of $\mathfrak{a}^\infty_m$. For a finite dimensional Lie group, the Lie algebra and the left invariant 1-forms (Maurer-Cartan forms) are in a natural way dual vector spaces. Moreover, the bracket operation in the Lie algebra and the exterior derivatives of the Maurer-Cartan forms determine each other by the formula

$$d\theta(X, Y) = -\theta([X, Y]).$$

With this operation d, the exterior algebra of the dual becomes a cochain complex. Since the procedure is purely algebraic, the dual of any finite dimensional Lie algebra is embedded into a cochain complex by these formulas. For an infinite dimensional Lie algebra, it is necessary to decide which topology one wants to use before introducing the dual, but once that is done, the procedure goes forward also. In the cases of interest at the moment, explicit discussion of the topology will be avoided by using direct and inverse limits of finite dimensional vector spaces. Indeed, the inverse system of epimorphisms by which $\mathfrak{a}_m^\infty$ is defined dualizes to a direct system of monomorphisms whose direct limit will be defined to be the dual space $(\mathfrak{a}_m^\infty)^*$ of the original Lie algebra. The following proposition then describes the method for making the exterior algebra of this dual space into a cochain complex.

1.16 Proposition. *There is a unique linear operator d on the exterior algebra $\Lambda(\mathfrak{a}_m^\infty)^*$ satisfying*

(i) $d1 = 0$

(ii) $d\theta(X, Y) = -\theta([X, Y])$ *for every* $\theta \in (\mathfrak{a}_m^\infty)^*$ *and* $X, Y \in \mathfrak{a}_m^\infty$.

(iii) $d(\phi \wedge \psi) = d\phi \wedge \psi + (-1)^r \phi \wedge d\psi$ *for every* $\phi \in \wedge^r(\mathfrak{a}_m^\infty)^*$ *and* $\psi \in \wedge(\mathfrak{a}_m^\infty)^*$.

Proof. Given any $\theta \in (\mathfrak{a}_m^\infty)^*$, there is a k such that θ vanishes on $\prod_{p=k}^{\infty} {}^p\mathfrak{a}_m$, that is, θ may be identified with an element of $(\mathfrak{a}_m^k)^*$. Moreover, $d\theta$ is then defined by (ii) and may be identified with an element of $(\mathfrak{a}_m^{k+1})^*$, thus with an element of $(\mathfrak{a}_m^\infty)^*$. Since $\Lambda(\mathfrak{a}_m^\infty)^*$ is generated as an algebra by 1 and $(\mathfrak{a}_m^\infty)^*$, there is at most one way to extend the operation defined by (i) and (ii) to all of the exterior algebra so that the product rule (iii) holds. For $\phi \in \Lambda^r(\mathfrak{a}_m^\infty)^*$, $d\phi$ must then be given by

$$d\phi(X_0, \ldots, X_r)$$
$$= \sum_{i<j} (-1)^{i+j} \phi([X_i, X_j], X_1, \ldots, X_{i-1}, X_{i+1}, \ldots, X_{j-1}, X_{j+1}, \ldots, X_r).$$

However, the latter formula can be used as a definition of d, so that its existence is proved. □

1.17 Corollary. *Let $\theta^i_{i_1 \ldots i_r}$ be the element of $(\mathfrak{a}_m^\infty)^*$ defined by*

$$\theta^i_{i_1 \ldots i_r}\left(x^{j_1} \ldots x^{j_s} \frac{\partial}{\partial x^j}\right) = \begin{cases} 1 & \text{if } i = j \text{ and } j_1 \ldots j_s \text{ is a permutation of } i_1 \ldots i_r \\ 0 & \text{otherwise.} \end{cases}$$

Then the set

$$\{\theta^i_{i_1 \ldots i_r} \mid i_1 \leqq i_2 \leqq \ldots \leqq i_r, 0 \leqq r < \infty\}$$

is a basis for $(\mathfrak{a}_m^\infty)^*$, and the exterior derivatives of these elements are given by

$$d\theta^i + \theta^\alpha \wedge \theta^i_\alpha = 0$$

$$d\theta^i_j + \theta^\alpha \wedge \theta^i_{\alpha j} + \theta^\alpha_j \wedge \theta^i_\alpha = 0$$

and for $r \geqq 2$

$$d\theta^i_{i_1 \ldots i_r} + \sum_{v=0}^{r} \sum \theta^\alpha_{j_1 \ldots j_v} \wedge \theta^i_{\alpha k_1 \ldots k_{r-v}} = 0$$

where the last sum is over all terms such that $\{i_1, \ldots, i_r\} = \{j_1, \ldots, j_v, k_1, \ldots, k_{r-v}\}$.

Proof. Since $\left\{x^{j_1}\dots x^{j_s}\frac{\partial}{\partial x^j}\right\}$ is a basis for $\mathfrak{a}_m^\infty$, $\{\theta^i_{i_1\dots i_r}\}$ is a basis for $(\mathfrak{a}_m^\infty)^*$. To prove the formulas for d, it suffices to verify that the value of the left side of each equation is 0 on each pair of basis vectors. This verification is a straightforward calculation. $\square$

1.18 Corollary. *On the subset of* $(\mathfrak{a}_m^k)^*$ *consisting of forms which vanish on the kernel of* Π^k_{k-1}, *the exterior derivative* d *is well-defined by the corresponding operation in* $(\mathfrak{a}_m^\infty)^*$. *In particular,* $d\theta^i_{i_1\dots i_r}$ *is defined for* $r \leqq k-1$, *but the above formulas hold on this subset only for* $r \leqq k-2$.

Proof. The first statement is clear from the proof of Proposition 1.16. The formula for $d\theta^i_{i_1\dots i_r}$ contains only $\theta^j_{j_1\dots j_s}$ with $s \leqq r$, except for the terms $\theta^\alpha \wedge \theta^i_{\alpha i_1\dots i_r}$. Thus, the formula for $d\theta^i_{i_1\dots i_{k-1}}$ makes no sense, but the other formulas make sense and are proved by the same calculations used in the preceding corollary.

2. Prolongation of a Manifold

The truncated polynomial groups GP^k_m introduced in the last section play the roll in nonlinear differential geometry that GL_m plays in linear differential geometry. In particular, each C^k manifold of dimension m has associated with it a principal GP^k_m-bundle called the k-jet bundle or the k-th prolongation of the manifold. Any diffeomorphism or vector field can be prolonged to jet bundles of appropriate orders. Just as a tangent vector to a manifold can be viewed as a derivation on the space of functions which depends only on the first order terms of the Taylor series, so a tangent vector to a jet bundle can be viewed as a derivation which depends on a finite number of terms of the Taylor series, that is, as a bounded derivation of the ring of formal power series. An equivalent statement is that the jet bundle admits a canonical 1-form with values in the space of derivations. This form is called the tautological form because it precisely establishes the amount of redundacy in the structure of the jet bundle. Thus, the lifts of diffeomorphisms of the original manifold can be distinguished from other bundle maps of the jet bundle by the property that they leave the tautological form unchanged. These ideas will be developed in this section, along with the notion of tangent group of a Lie group, which is useful in studying them. This prepares the way for the next section, in which the prolongation of G-structures will be studied.

2.1 Definition. If M is a C^l-manifold of dimension m, $0 \leqq l \leqq \infty$, P is a point of M, and $k \leqq l$, then a *k-jet* at P is an equivalence class of C^k embeddings of $(\mathbb{R}^m, 0)$ into (M, P), where two such embeddings f and g are equivalent if the composition $g^{-1} \circ f$ has the same derivatives up to order k at 0 as the identity map.

The derivatives referred to in Definition 2.1 are taken with respect to a fixed coordinate system in $\mathbb{R}^m$, but the equivalence relation does not depend upon the choice of the coordinate system. The set of all k-jets at all points of M is denoted by $J^k M$, and called the k-th prolongation of M. There are projections

$$\pi^j_k \colon J^j M \to J^k M$$

defined for $j \geqq k$ because the k-th equivalence relation is less restricitve than the j-th. π^k_0 is the projection of a principal bundle with group GP^k_m. To prove this statement, the essential step is the following lemma.

2.2 Lemma. *Let $f\colon (\mathbb{R}^m, 0) \to (\mathbb{R}^n, 0)$ and $g\colon \mathbb{R}^n \to \mathbb{R}^q$ be maps of class C^l and let r be an integer such that $r \leqq l$. Then the coefficients in the r-th order Taylor series at 0 for $g \circ f$ depend only on the coefficients in the r-th order Taylor series at 0 for g and for f, and the dependence is given by the rule for truncated composition of polynomials (Lemma 1.1).*

Proof. Choose coordinates $(x^1, \ldots, x^m)$ in $\mathbb{R}^m$, $(y^1, \ldots, y^n)$ in $\mathbb{R}^n$, and $(z^1, \ldots, z^q)$ in $\mathbb{R}^q$. Then the mappings have Taylor series at 0 given by

$$y^j = f^j(x) = A^j_i x^i + \ldots + A^j_{i_1 \ldots i_r} x^{i_1} \ldots x^{i_r} + o(|x|^r)$$

$$z^k = g^k(y) = B^k + B^k_j y^j + \ldots + B^k_{j_1 \ldots j_r} y^{j_1} \ldots y^{j_r} + o(|y|^r)$$

Substituting the first expression in the second yields the truncated composition, plus some terms which are $o(|x|^r)$ and some terms which are $o(|f(x)|^r)$. If the sum of the latter terms is denoted by $F(x)$, then

$$\frac{|F(x)|}{|x|^r} = \frac{|F(x)|}{|f(x)|^r} \cdot \frac{|f(x)|^r}{|x|^r}.$$

Since $f(x)$ approaches 0 as x does, the first factor approaches 0. Since the second factor remains bounded, the product approaches 0, which completes the proof of the lemma. □

From this lemma, it follows easily that GP^k_m acts freely and transitively on the right of each fiber of

$$\pi^k_0 \colon J^k \mathbb{R}^m \to \mathbb{R}^m,$$

and Euclidean translations act on the left of $J^k \mathbb{R}^m$ in such a way as to establish a natural product structure in this bundle. The structure of $J^k M$ for arbitrary M can then be determined.

2.3 Proposition. *$J^k M$ has a topology making it a locally trivial principal GP^k_m bundle over M. If M is a C^l manifold and $k \leqq l < \infty$, then $J^k M$ is a C^{l-k} manifold and each π^k_j is a C^{l-k} mapping. If M is a C^∞ manifold, then for each finite k, $J^k M$ is a C^∞ manifold and π^k_j is a C^∞ mapping. If $k \leqq l < \infty$, then any C^l diffeomorphism $f\colon M \to N$ induces a C^{l-k} diffeomorphism*

$$J^k f \colon J^k M \to J^k N$$

such that for every $P \in J^k M$ and $g \in GP^k_m$,

$$((J^k f) P) g = (J^k f)(Pg)$$

and for every j with $0 \leqq j \leqq k$

$$\pi_j^k \circ J^k f = J^i f \circ \pi_j^k.$$

If $l = \infty$, *then* $J^k f$ *is* C^∞.

$$J^k(f \circ g) = J^k f \circ J^k g$$

and if f is the identity map of M, then $J^k f$ *is the identity map of* $J^k M$. *The mapping*

$$\pi_{k-1}^k \colon J^k M \to J^{k-1} M$$

is a vector bundle which is the pullback by π_0^{k-1} *of the tensor bundle with fiber* $\mathbb{R}^m \otimes S^k(\mathbb{R}^m)^*$ *associated to the tangent bundle of M, where* S^k *is the symmetrized tensor power.*

Proof. See Kobayashi [1961]. □

A topological action of a group G on M does not necessarily induce an action of the group on $J^k M$, even if each element $g \in G$ acts in a C^k manner. In the latter case, a function $J^k MxG \to J^k M$ is defined, but in general is not continuous. In the case of a smooth vector field on M, the corresponding 1-parameter group does act on $J^k M$, and so defines a vector field on $J^k M$.

2.4 Proposition. *Let X be a* C^{k+l} *vector field on M, where* $l \geqq 0$. *Then there is a* C^l *vector field* $J^k X$ *on* $J^k M$ *which commutes with the fundamental vector fields of the* GP_m^k *principal bundle structure and satisfies*

$$(T_P \pi_j^k)(J^k X) = J^j X$$

for each P in $J^k M$. *The mapping which takes X to* $J^k X$ *is injective and linear, and it commutes with Lie brackets if* $l \geqq 1$.

Proof. X defines a C^{k+l} mapping F, so $J^k F_t$ is a C^l mapping whose derivative with respect to t is also C^l, as the formulas show. $J^k X$ commutes with the fundamental vector fields because $J^k F_t$ commutes with the action of GP_m^k. Since $\pi_j^k \circ (J^k F_t) = J^j F_t$, $J^k X$ satisfies the corresponding formula. The last sentence follows from the coordinate expression for $J^k X$, which is given by Arnold [1973, p. 255]. That Lie bracket is preserved can also be seen from the fact that the Lie bracket can be defined by means of local flows. This completes the proof. □

The tangent space to any manifold at any point can be identified with a set of derivations of functions defined in the neighborhood of the point. In the case of $J^k M$, a tangent vector should also be related to a derivation of the set of functions pulled back from M. In fact, it is related to $\mathfrak{a}_m^\infty$, and as a consequence, the purely algebraic Proposition 1.16 has an analog in the existence on $J^k M$ of certain canonical 1-forms whose exterior derivatives are related to the differential in the cochain complex $(\mathfrak{gp}_m^\infty)^*$. In order to define these forms, some preliminary remarks will be required.

If $x^{i_1}\ldots x^{i_r}\dfrac{\partial}{\partial x^i}$ is regarded as a vector field on $\mathbb{R}^m$, then by Proposition 2.4 it can be lifted to a vector field on $J^k\mathbb{R}^m$, thus in particular defines a vector belonging to $T_E J^k\mathbb{R}^m$ where E is the k-jet at the origin of the identity map. Let P be a point in $J^{k+1}M$, and let f be an embedding of $\mathbb{R}^m$ into M such that $J^{k+1}f(0) = P$. Then

$$T_E J^k f\colon T_E J^k\mathbb{R}^m \to T_{P'} J^k M,$$

where $J^k f(0) = P'$, is well-defined because for any diffeomorphism, $TJ^k f$ depends only on $J^{k+1}f$. Thus, the lift of $x^{i_1}\ldots x^{i_r}\dfrac{\partial}{\partial x^i}$ maps to a vector at P'. For $r > 0$, these vectors depend only on P', and in fact fit together to give a vector field $X_i^{i_1\ldots i_r}$ such that

$$\{X_i^{i_1\ldots i_r} \mid 1 \leqq r \leqq k,\ l_1 \leqq \ldots \leqq l_r\}$$

is a basis for the fundamental vector fields for the principal bundle structure of $J^k M$ over M. On the other hand, the image of $\partial/\partial x^i$ at P' depends on the choice of P, but this dependence affects only the components tangent to $N^k_{m,k-1}$. Thus, there is a well-defined section X_i in the quotient bundle of $TJ^k M$ by the tangent bundle to the fiber of the principal bundle

$$N^k_{m,k-1} \to J^k M \to J^{k-1} M.$$

The tangent bundle to this fiber may also be desribed as the kernel of $T\pi^k_{k-1}$.

2.5 Definition. $\theta^i_{i_1\ldots i_r}$, $0 \leqq r \leqq k-1$, is the 1-form on $J^k M$ defining by requiring that

$$\theta^i_{i_1\ldots i_r}(X_j^{j_1\ldots j_s})$$

equals 1 if $i = j$ and $\{i_1,\ldots,i_r\}$ is a permutation of $\{j_1,\ldots,j_s\}$, and equals 0 otherwise. The form $\theta^i_{i_1\ldots i_r}$ is called a *tautological form.*

$$\theta = \{\theta^i_{i_1\ldots i_r} \mid 0 \leqq r \leqq k-1\}$$

may be regarded as a form on $J^k M$ with values in $\mathfrak{a}^{k-1}_m$. Indeed, at any point, the mapping induced by θ on

$$TJ^k M/\mathrm{Ker}\ T\pi^k_{k-1}$$

is defined to take $X_i^{i_1\ldots i_r}$ to $x^{i_1}\ldots x^{i_r}(\partial/\partial x^i)$. θ will be referred to as *the* tautological form on $J^k M$.

2.6 Proposition. *Suppose M is C^l, $l \geqq k+1$. Then the sections $T\pi^k_{k-1}X_i^{i_1\ldots i_r}$, $r \leqq k-1$, give a C^{l-k-1} trivialization of the quotient bundle of $TJ^k M$ by the kernel of $T\pi^k_{k-1}$. $\theta^j_{j_1\ldots j_s}$ is a C^{l-k-1} 1-form which vanishes on the kernel of $T\pi^k_{k-1}$, and θ is an isomorphism on the quotient space at any point. If $f\colon M \to N$ is a diffeomorphism of class C^l, $l \geqq k+1$, then*

$$(TJ^k f)\,X_i^{i_1\ldots i_r} = X_i^{i_1\ldots i_r} \qquad r \leqq k$$

$$(T^* J^k f)\,\theta^i_{i_1\ldots i_r} = \theta^i_{i_1\ldots i_r} \qquad r \leqq k-1.$$

If $k \geqq j \geqq r+1$, *then*

$$(T^* \pi_j^k)\, \theta^i_{i_1 \dots i_r} = \theta^i_{i_1 \dots i_r}$$

If $l \geqq k+2$, *the forms* $d\theta^i_{i_1 \dots i_r}$, $0 \leqq r \leqq k-1$, *are defined and left fixed by* $T^*(J^k f)$, *while for* $0 \leqq r \leqq k-2$ *there are formulas for these exterior derivatives which are given by Corollary* 1.17.

Proof. See Kobayashi [1961]. □

The fact that X_i is not a vector field on $J^k M$ means that the tangent bundle to the fiber has no natural complement. A connection can be introduced to provide such a complement, but a certain degree of inconsistency with the natural structures is necessary. The definition of "connection" and related ideas will be discussed in § 4.

It is necessary to determine the action of GP^k_m on θ which is induced by its action on $J^k M$. For this purpose, it is convenient to recall some general facts about Lie group actions.

2.7 Proposition. *Given any Lie group* G, *its tangent bundle* TG *has a natural Lie group structure, which is a semi-direct product*

$$1 \to V \to TG \rightleftarrows G \to 1$$

where V *is the vector space underlying the Lie algebra* $\mathfrak{g}$ *of* G *and* G *acts on* V *by the adjoint action. If* $\{X_i\}$ *is a basis for* $\mathfrak{g}$ *and* ξ_i *is tangent to* X_i, *then the structure equations for* TG *are given by:*

$$[X_i, X_j] = C^k_{ij} X_k$$
$$[X_i, \xi_j] = C^k_{ij} \xi_k$$
$$[\xi_i, \xi_j] = 0.$$

Any C^{k+1} *action of* G *on a manifold* N *induces a* C^k *action of* TG *on* TN. *If a* C^{k+1} *mapping* $f\colon M \to N$ *commutes with actions of* G *on the two manifolds, then* Tf *commutes with the induced actions of* TG. *If furthermore* $k \geqq 1$ *and* N *is a principal* G*-bundle over* M, *then* TN *is a principal* TG *bundle over* TM. *Any homomorphism of Lie groups induces functorially a homomorphism of their tangent groups.*

Proof. See Bourbaki [Lie Groups and Lie Algebras, Chapter III, §§ 2–3]. □

2.8 Proposition. *If* f *is a polynomial representing an element* x *of* $J^k \mathbb{R}^m$, g *is a polynomial representing an element* y *of* GP^k_m, *and* g^{-1} *is a polynomial representing the inverse element* y^{-1}, *then there are well-defined linear mappings*

$$Ad_y\colon T'_E J^k \mathbb{R}^m \to T'_E J^k \mathbb{R}^m$$
$$L_y\colon T'_E J^k \mathbb{R}^m \to T'_y J^k \mathbb{R}^m$$

induced by the compositions $g \circ f \circ g^{-1}$ *and* $g \circ f$ *respectively. Here* T' *denotes the quotient of the tangent bundle by the kernel of* $T\pi^k_{k-1}$ *and* GP^k_m *is considered as a subset of* $J^k \mathbb{R}^m$. *These operations satisfy the identities*

$$L_y \circ Ad_{y^{-1}} = R_y$$
$$Ad_{yz} = Ad_y Ad_z.$$

There is a homomorphism Ad from GP_m^k *to the automorphism group of* $T_E' J^k \mathbb{R}^m$ *defined by*

$$Ad(y) \cdot X = Ad_y X.$$

Ad_y *is defined on* $\mathfrak{a}_m^{k-1}$ *by means of the identification between* $\mathfrak{a}_m^{k-1}$ *and* $T_E' J^k \mathbb{R}^m$. *Let X be tangent to* $J^k M$ *at x and Y to* GP_m^k *at y. Then*

$$\theta(Xy) = Ad_{y-1}\,\theta(X)$$
$$\theta(xY) = L_{y_1^{-1}} Y_1$$

where $y_1 = \pi_{k-1}^k y$ *and* $Y_1 = \pi_{k-1}^k Y$. *Here Xy and xY denote the result of operations of* TGP_m^k *on the right of* $T_E' J^k M$.

Proof. See Kobayashi [1961]. □

Since $J^k M$ is locally isomorphic to $J^k \mathbb{R}^m$, the effect of the group action on θ is given in general by these formulas.

The forms $\theta^i_{i_1 \dots i_r}$ are useful in characterizing those bundle maps of $J^k M$ which are lifts of diffeomorphisms.

2.9 Proposition. *A mapping F:* $J^k M \to J^k M$ *is the prolongation of a* C^l *diffeomorphism of M,* $l \geqq k+1$, *if and only if F is a diffeomorphism of class* C^{l-k} *which commutes with the natural operation of* GP_m^k *on* $J^k M$ *and satisfies*

$$(T^* F)\,\theta^i_{i_1 \dots i_r} = \theta^i_{i_1 \dots i_r}$$

for $r \leqq k-1$. *A vector field Y on* $J^k M$ *is the prolongation of a vector field X of class* C^l, $l \geqq k+1$, *on M if and only if Y is a vector field of class* C^{l-k} *which commutes with the fundamental vector fields and satisfies*

$$(\mathcal{L}_Y \theta^i_{i_1 \dots i_r})(X_j) = 0 \qquad r \leqq k-1.$$

Such a vector field Y satisfies $\mathcal{L}_Y \theta^i_{i_1 \dots i_r} = 0$.

Proof. It is clear that a prolongation of a diffeomorphism or a vector field has the indicated properties. To prove the converse, one proceeds by induction on k, using coordinates on $J^k M$ constructed from coordinates on M. □

2.10 Remark. If the condition that F be a bundle map is dropped, it is still possible to draw some conclusions. Consider the case $k = 1$. Then local calculations based on the condition $T^* F\theta^i = \theta^i$ show that in the neighborhood of any point of $J^1 M$, F has the form of the prolongation of a mapping f of M.

2.11 Remark. $J^k M$ is the quotient of $\mathrm{Diff}^l M$ by the subgroup $\mathrm{Diff}^l(M, P(k))$ consisting of diffeomorphism whose k-jet at $P \in M$ is the identity. From this point of view, the forms $\theta^i_{i_1 \dots i_r}$, $r \leqq k-1$, are precisely the 1-forms on $J^k M$ which are invariant by the left action of $\mathrm{Diff}^l M$ on its quotient $J^k M$.

The foundations of the theory of jet bundles were laid by Ehresmann [1951], while the tautological forms were discovered by Kobayashi [1961]. The name tautological form is due to Bott [1975], who used it in connection with the study of infinite Lie algebras. A further useful reference on the material of this section is the book of Kobayashi [1972, Chapter I]. All of these references are rather sketchy, perhaps because most of the history of jet bundles and trundated polynomial groups will occur in the future.

3. Higher Order Structures

In the first two sections of this chapter, the behaviour of higher order derivatives on a smooth manifold has been studied. It is now necessary to see how additional structures on the manifold are defined by or imply conditions on these derivatives. A common condition is the reduction of the group of the k-jet bundle to a closed subgroup G of GP_m^l. If the group of $J^l M$ can be reduced to G, then for $l < k$, the group of $J^k M$ can also be reduced to a subgroup of GP_m^k. By studying the case of the reduction of $J^l \mathbb{R}^m$ to $\mathbb{R}^m \times G$, that is, the standard G-structure on $\mathbb{R}^m$ as defined in I.1, these subgroups of GP_m^k for $l = 1$ will be found explicitly. Indeed, the set of k-jets of automorphisms leaving the origin fixed of the standard G-structure is a subgroup of GP_m^k which is an extension of G and whose action on $J^1 \mathbb{R}^m$ preserves the subset $\mathbb{R}^m \times G$. Hence, this subgroup will be denoted by $E^k G$ and called the k-th extension of G. Since the standard structure occurs in the tangent space at each point, $E^k G$ is the group required. After the details of this argument have been given, the relation of $E^k G$ to integrability will be discussed.

For the bundle reduction problem, it is important to study pairs consisting of a Lie group and a closed subgroup. For foliations, it is important to look instead at pairs of Lie algebras. Indeed, it turns out that the cohomology of pairs of infinite dimensional Lie algebras can be used to define characteristic classes of foliations. Therefore, this section ends by summarizing the facts in the finite dimensional case, in preparation for treating the infinite dimensional case in Chapter III.

Let G be a closed subgroup of GL_m and $\mathfrak{g}$ the corresponding subalgebra of $\mathfrak{gl}_m$. (It is important to notice that G is a specific subgroup, not a conjugacy class of subgroups.) Then, for any manifold M, the operation of GL_m on $J^1 M$ induces an operation of G on $J^1 M$. If M is parallelizable, any product structure $M \times GL_m$ on $J^1 M$ contains a subset $M \times G$ defined by the embedding of G in GL_m. Since $J^1 \mathbb{R}^m$ has a canonical product structure defined by translation in $\mathbb{R}^m$, the preceding remark is particularly useful in that case.

3.1 Definition. A *G-structure* on M is a subset $E(G,M)$ of $J^1 M$ such that G acts simply transitively on the set

$$E(G,x) = E(G,M) \cap (\pi_0^1)^{-1}(x)$$

for each $x \in M$. It is of *class* C^k if each point of M has a neighborhood on which there exists a C^{k-1} section of $J^1 M$ with image in $E(G,M)$. The *standard G-*

structure on $\mathbb{R}^m$ is the subset $\mathbb{R}^m \times G$ of $J^1 M$ defined by the canonical product structure on $J^1 \mathbb{R}^m$. Given G-structures on M and N, a C^k *isomorphism* of these structures is a C^k diffeomorphism $f: M \to N$ such that $J^1 f$ is a C^{k-1} diffeomorphism between the manifolds $E(G,M)$ and $E(G,N)$. A C^k *infinitesimal automorphism* of $E(G,M)$ is a C^k vector field defined an open subset of M such that the corresponding (possibly partially defined) flow consists of C^k isomorphisms of $E(G,M)$ defined over open subsets of M. A C^k G-structure $E(G,M)$ is *integrable* if each point of M has an open neighborhood U on which the induced G structure $E(G,U) = E(G,M) \cap (\pi_0^1)^{-1}(U)$ is C^k isomorphic to the G-structure induced on some open subset of $\mathbb{R}^m$ by the standard G-structure.

3.2 Proposition. *A C^k G-structure $E(G,M)$ is a closed C^{k-1} submanifold of $J^1 M$ which is a C^{k-1} principal G-bundle over M under the given action of G. $J^1 M$ is foliated of class C^{k-1} by the images of $E(G,M)$ under the elements of GL_m, and the fundamental vector fields corresponding to the subgroup G are tangent to the leaves of this foliation. The restrictions of the forms θ^i to the leaves of the foliation are independent. If f is a C^k isomorphism of G-structures, then $J^1 f$ induces a C^{k-1} map of G-bundles. f is a C^k automorphism of the standard G structure on an open subset of $\mathbb{R}^m$ if and only if in terms of standard coordinates f is given by C^k functions $f^i(x^1, \ldots, x^m)$ such that the matrix with entries $(\partial f^i/\partial x^j)(x)$ is an element of G for each x in the domain of f. X is a C^k infinitesimal automorphism of the standard G-structure on $\mathbb{R}^m$ if and only if each component X^i is a C^k function of linear coordinates such that the matrix $(\partial X^i/\partial x^j)(x)$ belongs to $\mathfrak{g}$ for each x in the domain of X. For a C^k G-structure $E(G,M)$ with $k \geqq 1$ the following conditions are equivalent:*

(i) *$E(G,M)$ is integrable.*

(ii) *M has an atlas of C^k coordinate neighborhoods $\{U_\mu\}$ with coordinates (x_μ^i) such that each frame*

$$\left\{\frac{\partial}{\partial x_\mu^1}, \ldots, \frac{\partial}{\partial x_\mu^m}\right\}$$

belongs to $E(G,M)$.

(iii) *M has an atlas of C^k coordinate neighborhoods $\{U_\mu\}$ with coordinates (x_μ^i) such that for $x \in U_\mu \cap U_\nu$, the matrix with entries*

$$(\partial x_\mu^i/\partial x_\nu^j)(x)$$

is an element of G.

(iv) *Each point of M is the origin of a C^k coordinate neighborhood U having coordinate (x^i) and containing an open neighborhood V such that*

$$V + V = \{(x^i + y^i)/x \in V, y \in V\}$$

is contained in U and each of the mappings t_a defined by $t_a(x^i) = (x^i + a^i)$ for $a \in V$ is a C^k isomorphism from $E(G,V)$ to $E(G, V + a)$.

Proof. It is convenient to begin by obtaining a coordinate description of an arbitrary G-structure on $\mathbb{R}^m$. Linear coordinates on $\mathbb{R}^m$ are obtained by choosing a basis $\{e_i\}$ and taking for any point t the coordinates t^i, where $t = t^i e_i$. The identification of $\mathbb{R}^m$ with its tangent space is made by identifying e_i with $\partial/\partial t^i$. To an element A of GL_m is associated the matrix (A^i_j) defined by $Ae_i = A^j_i e_j$, so that $(At)^j = t^i A^j_i$. For any manifold M, a coordinate description of the action of GL_m on $J^1 M$ is then obtained in the form

$$(x^i, x^i_j)(A^i_j) = (x^i, x^i_l A^l_j).$$

A G-structure on M is completely described by giving one point out of each set $E(G, x)$, since all the others are obtained from this one by the action of G. In particular, a G structure on $\mathbb{R}^m$ can be described by a mapping E: $\mathbb{R}^m \to GL_m$, such that

$$E(G, \mathbb{R}^m) = \{t^i, E^i_k(t) A^k_j \mid A \in G\}$$

Two such mappings E and $\tilde{E}$ describe the same G-structure if and only if there is a map B: $\mathbb{R}^m \to G$ such that for every $t \in \mathbb{R}^m$

$$\tilde{E}^i_k(t) = E^i_j(t) B^j_k(t).$$

The G-structure is of class C^k if the mapping E is of class C^{k-1}, and if E and $\tilde{E}$ are both C^{k-1}, then B will be C^{k-1}. Since the image of E is closed C^{k-1} submanifold of $J^1 \mathbb{R}^m$ transverse to the fibers, while G is embedded as a closed subset of the fiber over t by $A \mapsto (E^i_k(t) A^k_j)$, $E(G, \mathbb{R}^m)$ is a closed C^{k-1} submanifold of $J^1 \mathbb{R}^m$. Since for any diffeomorphism f: $M \to N$, $J^1 f$ is a bundle map, it follows that the image of $E(G, M)$ by $J^1 f$ is a G-structure $E(G, N)$, and that the restriction of $J^1 f$ to $E(G, M)$ is a G-bundle map. In particular, this remark applies if the image is a preassigned G-structure. For a manifold M with a G-structure $E(G, M)$ and an atlas $\{f_\mu: U_\mu \to V_\mu \subseteq \mathbb{R}^m\}$, $f_\mu(E(G, U_\mu))$ is a G-structure on V_μ, and the coordinate changes induce G-bundle isomorphisms. Hence, $E(G, M)$ is a closed C^{k-1} submanifold of $J^1 M$, which is the total space of a C^{k-1} principal G-bundle. Since G is a closed subgroup of GL_m, the cosets Gg for all g in GL_m form a foliation of GL_m such that $\mathfrak{g}$ is tangent to the leaves. From a local coordinate description of $J^1 M$ which is adapted to $E(G, M)$, it is clear that the mapping which takes $E^i_k(t) A^k_j$ to the group element A^k_j is a C^{k-1} submersion which defines a C^{k-1} foliation of an open set in $J^1 M$, by pulling back the coset foliation of GL_m. Since GL_m acts on all of $J^1 M$, this foliation is globally defined. Since $\mathfrak{g}$ is tangent to the leaves on G, and the local projection maps each fiber diffeomorphically onto GL_m, the corresponding fundamental vector fields are tangent to the foliation on $J^1 M$. The θ^i are independent on the leaves because their vanishing defines the fiber and the foliation is tranverse to the fiber. For any C^k isomorphism f of G-structures, $J^1 f$ induces a C^{k-1} map of G-bundles. If f is a C^k diffeomorphism defined on an open subset of $\mathbb{R}^m$, it takes the C^k G-structure defined by E into the C^k G-structure

$$\left\{(f(x^i), \frac{\partial f^i}{\partial x^k}(x) E^k_l(x) A^l_j)\right\}.$$

f is a C^k automorphism of a given G-structure if and only if there is a C^{k-1} map B: $\mathbb{R}^m \to G$ such that

$$E^i_l(f(x)) = \frac{\partial f^i}{\partial x^j}(x) E^j_k(x) B^k_l(x).$$

Consider now the standard G-structure on $\mathbb{R}^m$. For this structure the image of E lies in G so f is an automorphism if and only if $\partial f^i/\partial x^j$ is in G, as required. (For a general G-structure, one obtains that $(\partial f^i/\partial x^j)(x)$ belongs to a conjugate of G which depends only on x). Given a C^k vector field X on an open subset of $\mathbb{R}^m$, X will be an infinitesimal automorphism of the standard G-structure if and only if $J^1 X$ is tangent to $E(G,M)$ at all points of $E(G,M)$. $J^1 X$ has the coordinate expression

$$X^i(x)\,\partial/\partial x^i + \frac{\partial X^i}{\partial x^j}(x)\,x^j_k\,\frac{\partial}{\partial x^i_k}$$

where the second term is the left invariant vector field on $\{x\} \times GL_m$ having component $(\partial X^i/\partial x^j)(x)$ at the identity. Hence, X is an infinitesimal automorphism if and only if for all x, $(\partial X^i/\partial x^j)(x)$ is an element of $\mathfrak{g}$.

Consider now the question of integrability of G-structures. For the standard G-structure on $\mathbb{R}^m$, one may take $E^i_i(x) = \delta^j_i$, so in coordinates the structure becomes

$$\{(x^i, A^i_j)\}.$$

Suppose that f is a C^k isomorphism of a restriction of this structure onto the structure given on some coordinate neighborhood U in M, say

$$\{(y^i, E^i_k(y)\,A^k_j)\}.$$

f is a C^k isomorphism means that f is a C^k mapping and

$$(f^i(x), (\partial f^i/\partial x^k)(x)) \in E(G, U).$$

However, in the interpretation of the principal bundle as a set of frames, this point is

$$\frac{\partial f^i}{\partial x^k}(x)\frac{\partial}{\partial y^i} = \frac{\partial}{\partial x^k}$$

so the x^i form a coordinate system on U with

$$\left\{\frac{\partial}{\partial x^1}, \ldots, \frac{\partial}{\partial x^m}\right\} \in E(G, U).$$

This proves that (i) implies (ii). Now assume (ii), and consider the partial flow defined by the vector field $\partial/\partial x^i$. This is just the set of translations of the form

$$t_r(x^j) = (x^j + \delta^j_i r)$$

for small values of the real number r. Since these are analytic in terms of a C^k coordinate system, they are C^k on M. On the other hand, the frame

$$\left\{\frac{\partial}{\partial x^1}, \ldots, \frac{\partial}{\partial x^m}\right\}$$

in terms of coordinates (x^i, x^i_j) on $J^1 U$ is given by $x^i_j = \delta^i_j$, so the G-structure in the same coordinates is just

$$\{(x^i, A^i_j) \mid A \in G\}.$$

Since the 1-jet of a translation is an automorphism of this structure, the set of all sufficiently small translations has the required properties, so (ii) implies (iv). Assuming (iv) and using the fact that a coordinate translation induces the identity on the fiber of $J^1 \mathbb{R}^m$, a C^k isomorphism between $E(G, V)$ and a restriction of the standard G-structure is easily constructed. The fact that (ii) is equivalent to (iii) is clear from the formula

$$\frac{\partial}{\partial x^j_\mu} = \frac{\partial x^i_v}{\partial x^j_\mu} \frac{\partial}{\partial x^i_v}.$$

The proof of Proposition 3.2 is now complete. □

3.3 Remark. The foliation of GL_m by consets of G is defined by the left invariant forms which vanish on $\mathfrak{g}$. Locally, these can be pulled back to $J^1 M$, but the pullback depends on the product structure chosen, so the different pullbacks need not fit together to make a set of globally defined forms. The choice of a connection provides a way to define such forms globally (see § 4).

3.4 Remark. It follows immediately from this proposition that an integrable G-structure is the same thing as a $\mathscr{G}^k_m(G)$ structure, as defined in I.1. Hence, every translation is an automorphism of the standard G-structure on $\mathbb{R}^m$, as is every element of G, acting as a linear transformation on $\mathbb{R}^m$. A vector field X with $\partial X^i/\partial x^j$ a constant element of $\mathfrak{g}$ gives rise to a 1-parameter subgroup of the affine extension of G, and is an infinitesimal automorphism. In particular, if $\partial X^i/\partial x^j = 0$, one obtains a 1-parameter group of euclidean translations.

3.5 Definition. $E^k G$ is the set of k-jets of C^∞ automorphisms of the standard G-structure on $\mathbb{R}^m$ which leave 0 fixed. $\mathfrak{E}^k \mathfrak{g}$ is the set of k-jets at 0 of C^∞ infinitesimal automorphisms of the standard G structure on $\mathbb{R}^m$. $N^k_l G$ is the subset of $E^k G$ consisting of k-jets whose l-jet is the identity. $\mathfrak{n}^k_l \mathfrak{g}$ is the subset of $\mathfrak{E}^k \mathfrak{g}$ consisting of k-jets whose l-jet is 0. $\mathfrak{a}^k \mathfrak{g}$ is the direct sum $\mathfrak{E}^k \mathfrak{g} \oplus \mathbb{R}^m$.

3.6 Proposition. $E^k GL_m = GP^k_m$, $\mathfrak{E}^k \mathfrak{gl}_m = \mathfrak{gp}^k_m$, *and for every* G, $E^1 G = G$, $\mathfrak{E}^1 \mathfrak{g} = \mathfrak{g}$. $N^{k+1}_k G$ *is a closed subgroup of* GP^{k+1}_m *having Lie algebra* $\mathfrak{n}^{k+1}_k \mathfrak{g}$. *The exponential map is an isomorphism from the additive structure of the Lie algebra* $\mathfrak{n}^{k+1}_k \mathfrak{g}$ *to the group structure of* $N^{k+1}_k G$. *The action of a k-jet f on* $N^{k+1}_k G$ *depends only on* $\pi^k_1 f$, *and* $N^{k+1}_k G$ *with this action is isomorphic to the tensor space*

$$\mathbb{R}^m \otimes S^{k+1}(\mathbb{R}^m)^* \cap \mathfrak{g} \otimes S^k(\mathbb{R}^m)^*$$

where S^k *is the symmetrized tensor power and* $\mathfrak{g}$ *is identified with a subalgebra of* $\mathbb{R}^m \otimes (\mathbb{R}^m)^*$. $\mathfrak{n}^{k+1}_k \mathfrak{g}$ *is isomorphic to the tangent space of the tensor space* $N^{k+1}_k G$. *The polynomial representatives of the elements of* $N^{k+1}_k G$ *have the form*

$$f^i(x) = x^i + a^i_{i_1 \ldots i_{k+1}} x^{i_1} \ldots x^{i_{k+1}}$$

where for all b_i *in* $\mathbb{R}^m$

$$a^i_{jj_1 \ldots j_k} b^{j_1}_1 \ldots b^{j_k}_k \in \mathfrak{g}.$$

$E^k G$ is a closed subgroup of GP^k_m having the Lie algebra $\mathfrak{E}^k \mathfrak{g}$. The sequences

$$1 \to N_k^{k+1} G \to E^{k+1} G \to E^k G \to 1$$
$$0 \to \mathfrak{n}_k^{k+1} \mathfrak{g} \to \mathfrak{E}^{k+1} \mathfrak{g} \to \mathfrak{E}^k \mathfrak{g} \to 0$$

are exact. $\mathfrak{a}^k \mathfrak{g}$ admits a partially defined bracket operation induced by that of $\mathfrak{a}^k_m$.

Proof. Most of these statements are immediate consequences of the definitions and the results in § 2 about the structure of GP^k_m. For further discussion of $N_k^{k+1} G$, see Sternberg [1964]. This reference also contains some interesting examples. □

3.7 Remark. The word prolongation is used in the literature for almost any construction that involves higher derivatives, such as $J^k M$, $E^k G$, $J^k f$ and many more, including $N_k^{k+1} G$. Since for example $E^k G$ and $J^k G$ are in general distinct, the use of the word prolongation for $E^k G$ has been avoided here. It will instead be called the k-th extension of G.

Let G be a Lie group with Lie algebra $\mathfrak{g}$, let $X \in \mathfrak{g}$, and let X_x be the value of X at $x \in G$. Then the canonical form ω on G is the Lie-algebra valued 1-form defined by $\omega(X_x) = X$. It satisfies the equation

$$d\omega = -\omega \wedge \omega,$$

where d is the exterior derivative for Lie algebra valued forms. The form θ on $J^k M$ is a form with values in $\mathfrak{A}^{k-1}_m$ which satisfies

$$d(\pi^{k-1}_{k-2} \circ \theta) = -\theta \wedge \theta,$$

and its restriction to any fiber is $\pi^k_{k-1} \circ \omega$, where ω is the canonical form of fiber. The generalization of these relations to G-structures is given by the following proposition.

3.8 Proposition. *If M admits a G-structure of class C^{k+l}, then $J^k M$ admits a C^l reduction of its group to $E^k G$, that is to say, a closed submanifold $E^k(G, M)$ of class C^l which is a C^l principal $E^k G$ bundle over M under the action of the subgroup $E^k G$ induced by the action of GP^k_m on $J^k M$. $J^k M$ has a C^l foliation by the images of $E^k(G, M)$ under the action of GP^k_m. The fundamental vector fields corresponding to $\mathfrak{E}^k \mathfrak{g}$ are tangent to this foliation, and the forms θ^i are independent on the leaves. Suppose $\mathfrak{a}^{k-1} \mathfrak{g}$ admits an $Ad(E^k G)$ invariant complement $\mathfrak{m}$ in $\mathfrak{a}^{k-1}_m$ and let ω^k_G denote the canonical form on the fibers of $E^k(G, M)$. Then the restriction of θ to each fiber of $E^k(G, M)$ is equal to $\pi^k_{k-1} \omega^k_G$, and the restriction of θ to $E^k(G, M)$ is a form θ^k_G with values in $\mathfrak{E}^{k-1} \mathfrak{g}$ such that:*

(i) *If $X \in \mathfrak{E}^k \mathfrak{g}$ and X_x is the value at $x \in E^k(G, M)$ of the corresponding fundamental vector field, then*

$$\theta^k_G(X_x) = \pi^k_{k-1} X.$$

(ii) *For every X tangent to $E^k(G, M)$ and every $g \in E^k G$,*

$$\theta^k_G(TR_g X) = Ad(g^{-1}) \theta^k_G(X).$$

(iii) *If Y is tangent to $E^k G$ at g and $x \in E^k(G,M)$, then*

$$\theta_G^k(R_Y x) = L_{y_1^{-1}} Y_1$$

where $g_1 = \pi_{k-1}^k g$ and Y_1 is $\pi_{k-1}^k Y$.

(iv) $$d(\pi_{k-2}^{k-1} \circ \theta_G^k) = -\theta_G^k \wedge \theta_G^k.$$

If f is a C^{k+l} isomorphism of G-structures, then $J^k f$ induces on the submanifold $E^k(G,M)$ a C^l map of $E^k G$ bundles. A C^k vector field on X is an infinitesimal automorphism of $E(G,M)$ if and only if $J^r X$ is tangent to $E^r(G,M)$ for each $r \leqq k$. $E(G,M)$ is integrable if and only if $E^k(G,M)$ is locally isomorphic to the k-th extension of the standard G-structure on $\mathbb{R}^m$.

Proof. If M admits a G-structure, then the tangent space at each point has a distinguished element 0, and with respect to a suitable basis it has the standard G-structure of $\mathbb{R}^m$. Thus the automorphisms of $J^k M$ which preserve the structure form a subgroup conyugate to $E^k G$. The proof is a routine application of these remarks. □

Note that since $N_1^k G$ is connected, invariance under $Ad(E^k G)$ is equivalent to invariance under $Ad(\mathfrak{E}^k \mathfrak{g})$ and $Ad(g)$ for one element g of each connected component of G. Furthermore, it is sufficient to find a complement to $\mathfrak{E}^{k-1} \mathfrak{g}$ in $\mathfrak{gp}_m^{k-1}$, and for such a complement it is sufficient to require invariance under $Ad(E^{k-1} G)$.

In general, there do exist G-structures which are not integrable. In the case of the orthogonal group O_m, any m-manifold admits a Riemannian metric, that is to say, an O_m-structure. On the other hand, to be integrable it must be locally isometric to euclidean space, which is a very stringent restriction. Indeed, if such a manifold is complete, it admits euclidean space as its universal covering. By contrast, the natural GL_m structure of a C^∞ manifold is always integrable. In general, the determination of when a G-structure is integrable, or more generally, when two structures are locally equivalent, is not easy. Information can sometimes be obtained in terms of various tensors on the space $E^k(G,M)$, but the nature of the problem depends heavily on G. For more information on this question see Sternberg [1964, p. 314 et seq.]. The particular case of foliations will be dealt with later. From another point of view, integrability is a question of admitting sufficiently many automorphisms, since it can be shown that the existence of a transitive abelian group of automorphisms implies integrability for a G-structure, even in the C^0 case (see Theorem I.4.12). The commutativity is essential, as is shown by the example of a plane field defined by a subspace of a Lie algebra which is not a subalgebra.

There also exist reductions of the group of $J^k M$ other than those induced by reductions of $J^1 M$. So far, these have not been much studied. It appears likely that such studies will be more frequent in the future, as part of the process of understanding nonlinear phenomena in physics and elsewhere.

In the first part of this section, various facts about reductions of the group of $J^k M$ have been collected. It is useful to add some general facts about pairs consisting of a Lie group and a subgroup, or a Lie algebra and a subalgebra. The following proposition is needed both here and in various other contexts.

3.9 Proposition. *Let $p\colon E \to M$ be a smooth fibration with a connected fiber F. Then the exterior algebra of differential forms on M is mapped isomorphically by p^* onto the algebra of forms on E such that*

(i) *$\omega(X_1, \ldots, X_q) = 0$ whenever ω is a q-form and at least one X_i is tangent to a fiber.*
(ii) *$d\omega(X_1, \ldots, X_{q+1}) = 0$ under the same conditions.*

The second condition can be replaced by

(ii′) *$\mathscr{L}_X \omega = 0$ whenever X is tangent to a fiber.*

Proof. Let ω be a form on M. Then

$$p^* \omega(X_1, \ldots, X_q) = \omega(p_* X_1, \ldots, p_* X_q)$$

and $p_* X_i = 0$ if X_i is tangent to a fiber. Since $dp^* = p^* d$, every pulled-back form satisfies the given condition, whether or not the fiber is connected. Conversely, suppose ω satisfies these two conditions, and introduce local coordinates (x^a, y^α) such that y^α are functions on M. Then by condition (i)

$$\omega = \sum \omega_{j_1 \ldots j_\alpha}(x, y)\, dy^{j_1} \wedge \ldots \wedge dy^{j_\alpha}$$

while by condition (ii), $\partial \omega_{j_1 \ldots j_k} / \partial x^a = 0$ for all a at every point in the coordinate neighborhood. Since F is connected, ω is the pullback of a form on M. If the conditions (i) and (ii) hold, then

$$\mathscr{L}_X \omega = i(X)\, d\omega + d i(X)\omega = 0.$$

and the same equation shows that (i) and (ii′) imply (ii). □

3.10 Corollary. *Let H be a closed subgroup of the Lie group G, and let $\mathfrak{h}$ and $\mathfrak{g}$ be their respective Lie algebras. Then the exterior algebra of differential forms on $Q = G/H$ which are invariant under the action of G is isomorphic by an isomorphism commuting with the exterior derivative d to the exterior algebra generated by the elements of $\Lambda \mathfrak{g}^*$ such that*

(i) *$\omega(X_1, \ldots, X_q) = 0$ whenever $\omega \in \Lambda^q \mathfrak{g}^*$ and at least one $X_i \in \mathfrak{h}$.*
(ii) *$\omega(Ad h X_1, \ldots, Ad h X_q) = \omega(X_1, \ldots, X_q)$ whenever $\omega \in \Lambda^q \mathfrak{g}^*$ and $h \in H$.*

In particular, the latter set is closed under d.

Proof. For left invariant vector fields and forms condition (ii′) of the proposition can be written

$$\operatorname{ad}(X)\omega = 0.$$

If H is connected, this is equivalent to condition (ii) of the corollary. If H is not connected, then (ii) can be proved by a direct calculation based on the following formula for the right action of TH on TG:

$$(g, X)(h, Y) = (gh, \operatorname{ad}(h^{-1})X + Y). \quad \square$$

If $\mathfrak{h}$ is a subalgebra of a finite dimensional Lie algebra $\mathfrak{g}$, then the corresponding group G is foliated by cosets of the connected group H corresponding to $\mathfrak{h}$, whether or not H is closed. The left invariant forms on G satisfying

$$i(X)\omega = \mathrm{ad}(X)\omega = 0$$

for every $X \in \mathfrak{h}$ then can be considered as left invariant forms on the quotient, even though the quotient may be highly degenerate as a topological space. Furthermore, for left invariant forms the exterior derivative is given purely algebraically, the formula for 1-forms being

$$d\omega(X, Y) = -\,\omega([X, Y]).$$

Thus, corresponding to the Lie algebra $\mathfrak{g}$, there is a cohomology algebra $H^*(\mathfrak{g})$ defined by considering the operator d on $\Lambda\ \mathfrak{g}^*$. Since the subspaces of $\Lambda\ \mathfrak{g}^*$ invariant by ad $\mathfrak{h}$ or $Ad\,H$ are also invariant by d, relative cohomology $H^*(\mathfrak{g}, \mathfrak{h})$ and $H^*(\mathfrak{g}, H)$ is defined. These cohomology theories may or may not be related to the cohomology of G (respectively G/H) as topological spaces. If G is compact connected, $H^*(\mathfrak{g})$ is isomorphic to $H^*(G)$, so in particular if $\mathfrak{g}$ is abelian $H^*(\mathfrak{g})$ is isomorphic to the cohomology of a torus. If G is simple, $H^*(\mathfrak{g})$ is isomorphic to the cohomology of the corresponding compact real form. If K is the maximal compact subgroup of G, then the pair consisting of the complexes for $\mathfrak{g}$ and for $(\mathfrak{g}, K)$ gives rise to a spectral sequence in which E_2 is

$$H^r(K) \otimes H^s(\mathfrak{g}, K)$$

and the limit is $H^*(\mathfrak{g})$. (Since G/K is contractible, $H^r(K)$ is isomorphic to $H^r(G)$.)

If K is the identity, this sequence yields no information, and it is necessary to enlarge the complex to find an interpretation for $H^*(\mathfrak{g})$. In fact, one must take account of the cohomology of the classifying space BG of G. (This is known to algebraists as the cohomology of G, but it is not the cohomology of the topological space of G.) To do this, one constructs an algebraic analog of the fibration

$$TG \to (TG \times EG)/G \to EG/G = BG$$

where EG is the total space of the classifying bundle for G. The cochain complex consists of smooth functions f on $\mathfrak{g}^r \times G^{s+1}$ which are alternating r-linear in the first r variables. Then

$$\phi_y(x, L_x X_1, \ldots, L_x X_r) = f(X_1, \ldots, X_r, x, y^1, \ldots, y^s)$$

may be regarded as a form on G parametrized by G^s. The coboundary is a combination of the exterior derivative and the coboundary in the algebraic construction for BG. This complex gives rise to a spectral sequence converging to $H^*(\mathfrak{g})$ with E_2 term

$$H^r(G) \otimes H^s_{\mathrm{smooth}}(BG)$$

where $H^*_{\mathrm{smooth}}(BG)$ is defined by taking $r = 0$ in the preceding construction. This spectral sequence can be generalized to calculate $H^*(\mathfrak{g}, K)$ for compact K, in which case G is replaced by G/K in the E_2 term. In particular, if K is the maximal compact subgroup, it follows that $H^*(\mathfrak{g}, K)$ is isomorphic to $H^*_{\mathrm{smooth}}(BG)$. These spectral sequences are due to van Est [1953, 1955, 1958]. Since in these results, $H^*(G)$ appears as the de Rham cohomology and therefore also as a smooth cohomology theory, one can say that the cohomology of a finite dimensional Lie algebra is related to the smooth cohomology of appropriate spaces.

The infinite dimensional Lie algebra $\mathfrak{a}_m^\infty$ and various subalgebras are important in the theory of G-structures. The cohomology of these algebras, as well as the cohomology modulo compact subgroups of GL_m, is defined as above, except that cochains must be required to be continuous with respect to the inverse limit topology. This cohomology will turn out to give characteristic classes for foliations, by reason of the fact that it can be related to the smooth cohomology of certain topological spaces which arise as classifying spaces for foliations. Since the classifying spaces involved actually are related to singular foliations of a certain kind, further discussion will be deferred until Chapter III.

The theory of G-structures grew out of the local theory of Lie and Cartan discussed in Chapter I, dealing with transformations defined by systems of partial differential equations. Its formulation in terms of bundles and related global concepts is due to Ehresmann [1952 and later], while Chern [1953] studied the partial differential equations from the bundle theoretic point of view. Some more recent developments are discussed by Sternberg [1964], Guillemin and Sternberg [1966], and Kumpera and Spencer [1972]. The point of view presented here is close to that of Ehresmann in his early papers on this subject.

4. Connections and Characteristic Classes

A general theory of connections in arbitrary smooth principal bundles exists and has been adequately presented elsewhere, for example, by Kobayashi and Nomizu [1963, 1969]. The aim of this section is to present the special features of connections in subbundles and quotient bundles of $J^k M$, in order to apply the theory to the study of G-structures and their extensions. This has the additional advantage of aiding in the understanding of the general theory of connections by means of the comparison and contrast between the connection form and the tautological form. The section begins with a definition of connection and some remarks on aspects of the general theory most relevant for the study of characteristic classes of foliations (see Chapter III). Then linear and affine connections are reviewed briefly, and finally connections in $J^k M$ are discussed.

Let $E = E(G, M)$ be the total space of a C^k principal G bundle over M, $k > l$, with projection $\pi\colon E \to M$. Let $\mathfrak{g}$ be the Lie algebra of G.

4.1 Definition. A C^l *connection* in E is a homomorphism σ from $\mathfrak{g}^*$ to the C^l 1-forms on E such that

(i) For every $x \in E$ and $X \in \mathfrak{g}$,

$$\sigma(\phi)(X_x) = \phi(X)$$

where X_x is the value at x of the fundamental vector field corresponding to X and ϕ belongs to $\mathfrak{g}^*$.

(ii) For every $g \in G$ and X tangent to E

$$\sigma(\phi)(TR_g X) = \sigma(Ad(g^{-1})\phi)(X).$$

The equivalence of this definition with those given by Kobayashi and Nomizu [1963, pp. 63–68] follows easily from the formula

$$\sigma(\phi)(X) = \phi(\omega(X))$$

where ω is the connection form.

Note that the curvature of a connection vanishes identicaly if and only if the horizontal spaces are tangent to a foliation. Many nontrivial examples of foliations arise in this way. Thus, the theory of flat bundles provides suggestions for both theorems and proofs in foliation theory.

For the study of characteristic classes, it is important to understand the relation of curvature to the homomorphism σ. Though σ extends immediately to a homomorphism from the exterior algebra $\wedge\, \mathfrak{g}^*$ to the differential forms on E, this homomorphism fails to commute with the exterior derivative, and the failure is measured by the curvature. It is convenient to introduce additional algebraic tools before deriving the precise formula that describes this failure and its most important consequences. These tools are algebraic analogs of well-known operations in differential geometry.

Let $\mathfrak{g}$ be a finite dimensional Lie algebra with basis $\{e_i\}$ and let $\{e^j\}$ be the dual basis for $\mathfrak{g}^*$. An antiderivation of the exterior algebra $\Lambda\mathfrak{g}^*$ is an endomorphism D of the linear structure such that

$$D(\alpha \wedge \beta) = D\alpha \wedge \beta + (-1)^p \alpha \wedge D\beta$$

provided $\alpha \in \Lambda^p\, \mathfrak{g}^*$. There are two important types of antiderivations: the exterior derivative, and the interior product by a vector of $\mathfrak{g}$. The latter are specified uniquely by the conditions

$$i(e_i)r = 0 \qquad r \in \mathbb{R}$$
$$i(e_i)e^j = \delta_i^j$$

and the requirements that $i(X)$ be linear in X, and an antiderivation for each fixed X. Out of these, operators $L(X)$ corresponding to the Lie derivative are constructed by defining

$$L(X) = i(X)d + di(X).$$

$L(X)$ is a derivation, in the sense that it is an endomorphism of $\Lambda\mathfrak{g}^*$ which satisfies

$$L(X)(\alpha \wedge \beta) = L(X)\alpha \wedge \beta + \alpha \wedge L(X)\beta.$$

It is easy to verify that d then satisfies

$$d = \tfrac{1}{4} e^l \wedge L(e_l).$$

4.2 Lemma. *Let ω be a connection form on E and let σ be the corresponding homomorphism defined on $\mathfrak{g}^*$, and also its extension to $\Lambda\mathfrak{g}^*$. Then for every $\phi \in \mathfrak{g}^*$*

$$d(\sigma(\phi)) - \sigma(d\phi) = \phi(\Omega)$$

where Ω is the curvature form of the connection.

Proof. σ and ω are related by the formulas

$$\sigma(\phi)(X) = \phi(\omega(X)).$$

It is sufficient to consider the case $\phi = e^k$, and to evaluate both sides of the formula on a pair (X, Y), where X and Y vary over all possible combinations of horizontal and fundamental vector fields. Each side is then 0, except in the case that X and Y are both horizontal. In that case, the equation becomes

$$-e^k(\omega([X, Y])) = e^k(\Omega(X, Y))$$

which is a known result of connection theory. □

The relation between curvature and characteristic classes arises from the fact that certain forms constructed with the aid of the curvature can be projected to the base manifold where they define cohomology classes which do not depend on the connection. On the other hand, connections and curvatures behave sufficiently nicely with respect to bundle mappings that these cohomology classes can be identified with the characteristic classes of the bundle. A precise statement of the behavior of connections under G-bundle mappings is contained in the following proposition.

4.3 Proposition. *Let $E_i(G, M_i)$ be principal bundles with group G over M_i and let $f: M_1 \to M_2$ be a C^k mapping such that $f(xg) = f(x)g$ for every $g \in G$ and $x \in M_1$. Consider connections in E_i with form ω_i, curvature Ω_i, and homomorphism σ_i. Then, given any connection in E_2, there is a unique connection in E_1 such that $(T^*f)\omega_2 = \omega_1$, $(T^*f)\Omega_2 = \Omega_1$, and $T^*f \circ \sigma_2 = \sigma_1$.*

Proof. Define ω_1 by $\omega_1 = (T^*f)\omega_2$, and verify the remaining details. □

It is clear that to apply this proposition to study cohomology classes constructed on an arbitrary manifold, it is enough that the classes be independent of the choice of connection, and that for connections related as in the proposition, the actual forms be mapped by T^*f. How can such forms be constructed? What are needed are real valued forms on the base space. The pullbacks of such forms to the total space can be characterized as horizontal forms which are invariant by the action of the group (compare Proposition 3.9 and Corollary 3.10). The curvature form meets one of these conditions, since it is horizontal. However, it is Lie algebra valued, and is modified by the adjoint action when an element of the group acts on the bundle. Thus, if ϕ is a real valued function on $\mathfrak{g}$ which is invariant under the adjoint action, then $\phi \circ \Omega$ is a form on the total space which can be projected to the base. In fact, it is necessary to consider functions on the Cartesian powers $\mathfrak{g}^p$. The p-multilinear, symmetric functions on $\mathfrak{g}$ which are invariant under the adjoint action of $\mathfrak{g}$ give rise to closed (by the Bianchi identity) $2p$-forms on the base space, which have the required properties (see Proposition 4.4 below). Since for any Lie group G and any dimension N, there is an analytic manifold and an analytic G-bundle over it which serves as a classifying space for G-bundles over manifolds of dimension at most N, the above construction can be applied to this N-classifying bundle. If G is compact, the Lie algebra construction produces exactly the cohomology of the base space, at least up to dimension N (see H. Cartan [1949/50, exp. 20]). The pullback of this cohomology by the classifying map of any smooth bundle is therefore both the characteristic classes and the classes obtained by applying the construction directly to the given bundle. The complete program will not be carried out here (see H. Cartan [1949/50] and Koszul [1950]), but it is necessary to examine closely the process by which these differential forms are produced on the base space of a bundle with a connection, and to extend σ to a homomorphism on a larger algebra which copies this process.

4.4 Proposition. *Let G be a Lie group with Lie algebra $\mathfrak{g}$, and let ϕ be a function of p arguments, each belonging to $\mathfrak{g}$, which is multilinear, symmetric, and satisfies*

$$\phi(Ad(g)X_1, \ldots, Ad(g)X_p) = \phi(X_1, \ldots, X_p)$$

for every $g \in G$. Let ω be a connection form of a G bundle E and Ω the corresponding curvature form. Then $\phi(\Omega, \ldots, \Omega)$ is a closed $2p$-form on E which projects to a closed $2p$-form on M. On E, $\phi(\Omega, \ldots, \Omega) = d\phi(\omega, \Omega, \ldots, \Omega)$. Given two connections ω_0 and ω_1 in the same bundle E over M, then

$$\omega_{01} = (1-t)\omega_0 + t\omega_1$$

is a connection in the bundle $E \times \mathbb{R}$ over $M \times \mathbb{R}$ with curvature Ω_{01}. There is an operation A taking p-forms on $M \times \mathbb{R}$ to $(p-1)$-forms on M so that

$$d(A(\phi(\Omega_{01}, \ldots, \Omega_{01}))) = \phi(\Omega_1) - \phi(\Omega_0).$$

Proof. See for example Kobayashi and Nomizu [1969, pp. 293–298]. □

It is now clear why it is useful to extend the definition of the map σ (Definition 4.1) to include consideration of the curvature of the connection. Recall (Lemma 4.2) that the natural extension of σ to Λ^* satisfies

$$d\sigma(\phi) - \sigma(d\phi) = \phi(\Omega)$$

for every $\phi \in \mathfrak{g}^*$. Since $\phi(\Omega)$ is a 2-form, $\phi(\Omega)$ commutes with $\psi(\Omega)$ for every $\phi, \psi \in \mathfrak{g}^*$. Thus, it is reasonable to introduce the symmetric algebra $S(\mathfrak{g}^*)$, and to try to extend both d and σ to the algebra

$$W(\mathfrak{g}) = \Lambda\mathfrak{g}^* \otimes S\mathfrak{g}^*$$

in such a way that σ commutes with d. Assume for the moment that G is connected, so the adjoint action of G in $\mathfrak{g}^*$ is determined by the adjoint action of $\mathfrak{g}$. It is useful to extend the operators $i(X)$ and $L(X)$ (introduced in connection with Lemma 4.2) to $W(\mathfrak{g})$ in a manner consistent with their differential geometric analogs, so that σ will commute with them also. When an element ϕ of $\mathfrak{g}^*$ is viewed as an element of $S\mathfrak{g}^*$, it will be denoted by the corresponding capital letter Φ. $W(\mathfrak{g})$ is made into a graded algebra by taking ϕ of degree 1 and Φ of degree 2, and it is graded commutative (the change of signs when the order of factors is reversed corresponds to the behavior of differential forms of the same degree). Since $\phi(\Omega)$ is horizontal, one defines $i(X)\Phi = 0$ and then extends $i(X)$ to $W(\mathfrak{g})$ as an antiderivation. Let h be the isomorphism that takes ϕ to Φ. Then $L(X)$ is the derivation which satisfies $L(X)\Phi = h(L(X)\phi)$, that is, it does the same thing to ϕ in whichever part of the algebra it is viewed. (In fact, in either case it expresses the adjoint action of $\mathfrak{g}$ in $\mathfrak{g}^*$.) An element α of $W(\mathfrak{g})$ is called basic if $i(X)\alpha = L(X)\alpha = 0$ for every $X \in \mathfrak{g}$.

In the two preceding definitions, $i(X)$ and $L(X)$ are extensions of the operations already given in $\Lambda\mathfrak{g}^*$. But d must be redefined on $\Lambda\mathfrak{g}^*$ as well, to take account of Lemma 4.2. Call the new operator δ. Then δ must satisfy

$$\delta\phi = d\phi + \Phi,$$

since $\sigma(\Phi)$ should be defined to be $\phi \circ \Omega$. Also, since $i(X)\,d\sigma(\Phi) = L(X)\,\sigma(\Phi)$, it is natural to require that $i(X)\,\delta\Phi = L(X)\,\Phi$. With this background it is now possible to state the principal properties of the algebra $W(\mathfrak{g})$.

4.5 Proposition. *Let G be a connected Lie group with Lie algebra $\mathfrak{g}$, and let σ be a homomorphism from $\mathfrak{g}^*$ to the C^l 1-forms on the principal bundle $E(G, M)$ which is a connection in this bundle. Then there is a unique antiderivation δ in the algebra $W(\mathfrak{g})$ such that*

$$\delta\phi = d\phi + \Phi, \qquad i(X)\,\delta\Phi = L(X)\,\Phi.$$

There is a unique algebra homomorphism from $W(\mathfrak{g})$ to the differential forms on $E(G, M)$ such that $\sigma(\phi) = \phi \circ \omega$ and $\sigma(\Phi) = \phi \circ \Omega$, where ω and Ω are the connection and curvature forms of σ. σ commutes with $i(X)$ and $L(X)$ for every $X \in \mathfrak{g}$, and $\sigma \circ \delta = d \circ \sigma$. The cohomology of $W(\mathfrak{g})$ under δ is trivial. If G is compact and connected, the cohomology of the basic elements of $W(\mathfrak{g})$ is isomorphic to the cohomology of the classifying space for G, the cohomology of $\Lambda\mathfrak{g}^*$ under d is isomorphic to the cohomology of G as a topological space, and the filtration of $W(\mathfrak{g})$ by the degree of $S\mathfrak{g}^*$ gives rise to a spectral sequence isomorphic to the spectral sequence of the universal bundle. Hence, the image of σ is the same cohomologically as the image of the spectral sequence of the universal bundle, and in particular, the basic elements of the image give the characteristic classes of the bundle. The basic elements of $W(\mathfrak{g})$ can be identified with the $Ad(\mathfrak{g})$-invariant polynomials on $\mathfrak{g}$.

Proof. All these results are contained in the papers of H. Cartan and Koszul cited above. The homomorphism σ is called the Chern-Weil homomorphism. □

4.6 Remark. It is important to observe that the forms $\phi(\Omega, \ldots, \Omega)$ are defined and may very well be nonzero even if the cohomology classes they define are 0. They and the forms $\phi(\omega, \Omega, \ldots, \Omega)$ have uses that go beyond the elementary theory of characteristic classes. Indeed, much of the recent progress in foliation theory has been motivated by the observation that in some cases, some of these forms vanish, and that this fact enables one to make useful combinations of the other forms.

The calculation of these so-called invariant polynomials ϕ is a purely algebraic task. Some important special cases are found in Kobayashi and Nomizu [1969, pp. 298–305]. For the general linear group GL_m, the invariant polynomials themselves form a polynomial algebra over the reals with generators C_i, $i = 0, 1, \ldots, m$, where C_i is a polynomial of degree i. Thus, the forms $C_i(\Omega)$ exist for all i, but for odd i they are always exact. The classes given by $C_{2i}(\Omega)$ are called the Pontrjagen classes of the bundle. Since the injection map of the orthogonal group O_m into GL_m is a homotopy equivalence, the two groups have the same characteristic classes. But O_m has fewer invariant polynomials, since the restriction to $\mathfrak{o}_m$ of C_i vanishes for odd i, and no new invariant polynomials exist on the smaller algebra. On the other hand, the connected group SO_m does admit more invariant polynomials if m is even, namely, products involving the polynomial corresponding to the Euler class. Thus, the list of real characteristic classes for an SO_m (or GL_m^+) bundle is completed. On the other hand, as has already been remarked in § 3, GL_m and O_m are very different in the context of integrability problems for G-structures. Thus, it is important to have the result for GL_m, even though from a purely topological point of view it is not essential.

This completes the review of properties of connections in arbitrary G bundles. In the case that G is a closed subgroup of the general linear group, any G bundle has an associated principal affine bundle. Indeed, there is a subgroup $A_m(G)$ of the affine group and a split exact sequence

$$1 \to \mathbb{R}^m \xrightarrow{\alpha} A_m(G) \underset{\gamma}{\overset{\beta}{\rightleftarrows}} G \to 1.$$

For any given G-bundle, there is a corresponding sequence of bundles over the same base space, and for any connection $\tilde{\omega}$ in the affine bundle

$$\gamma^* \tilde{\omega} = \omega + \phi$$

where ω is a connection in the G-bundle and ϕ is a horizontal, invariant $\mathbb{R}^n$-valued form on the G-bundle. Conversely, any such pair of forms on the G-bundle gives rise to a connection in the $A_m(G)$-bundle. Also

$$\begin{aligned} \gamma^* \tilde{\Omega} &= \Omega + D\phi \\ d\phi &= -\,\omega \wedge \phi + D\phi \\ d\omega &= -\,\omega \wedge \omega + \Omega. \end{aligned}$$

The preceding remarks are slight generalizations of those of Kobayashi and Nomizu [1963, pp. 125–130] and are proved by the same methods. In the case of a G-structure, it is natural to choose ϕ to be the restriction of the tautological form θ. Then $\Theta = D\theta$ is called the torsion of the G-connection, and the preceding equations become

$$\begin{aligned} \gamma^* \tilde{\omega} &= \omega + \theta \\ \gamma^* \tilde{\Omega} &= \Omega + \Theta \\ d\theta &= -\,\omega \wedge \theta + \Theta \\ d\omega &= -\,\omega \wedge \omega + \Omega. \end{aligned}$$

In the case of $J^1 M$, the three bundles in the exact sequence are the tangent bundle, the bundle of affine frames, and the bundle of linear frames.

Consider now $J^k M$, equipped with the connection from ω for a GP_m^k connection and the tautological form θ. These have components

$$\begin{aligned} &\omega^i_{i_1 \ldots i_r} \qquad 1 \leqq r \leqq k \\ &\theta^i_{i_1 \ldots i_r} \qquad 0 \leqq r \leqq k-1, \end{aligned}$$

and satisfy the structure equations

$$\begin{aligned} d\omega + \omega \wedge \omega &= \Omega \\ d(\pi^{k-1}_{k-2}\, \theta) + \theta \wedge \theta &= 0. \end{aligned}$$

Both forms have the same value on a fundamental vector field for GP_m^{k-1}, and they are affected in the same way by the right action of GP_m^k. In spite of these similarities, one cannot hope to find a connection with $\omega^i_{i_1 \ldots i_r} = \theta^i_{i_1 \ldots i_r}$ for $1 \leqq r \leqq k-1$. Indeed, the equation

$$d\omega^i_j + \omega^\alpha_j \wedge \omega^i_\alpha = \Omega^i_j$$

shows that $d\omega^i_j$ vanishes on a pair consisting of a horizontal and a vertical vector. On the other hand, the equation

$$d\theta^i_j + \theta^\alpha \wedge \theta^i_{\alpha j} + \theta^\alpha_j \wedge \theta^i_\alpha = 0,$$

which holds for $k \geqq 3$, shows that there are such pairs for which $d\theta^i_j$ does not vanish. Therefore, one must seek a weaker relation between θ and ω. In some cases, a satisfactory answer is provided by the notion of Cartan connection (Kobayashi [1972, pp. 127–138]). But in general, there is no known fully satisfactory theory of connections in subbundles of $J^k M$.

To attempt even a brief history of the ideas summarized in this section is inadvisable. For details and some historical information, the reader may consult the books of Kobayashi and Nomizu [1963, 1969] and Husemoller [1966], and the memoir of M. Mostow [1976].

5. Foliations, Connections, and Secondary Classes

In the preceding sections of this chapter, the general theory of G-structures and their extensions has been developed, including the theory of connections appropriate to such structures. In this section, the groups G associated to foliations will be studied, and the integrability problem for such G-structures will be related to the more direct study of the integrability problem for plane fields which is given in I.4. In particular, the additional structure provided by the connection makes it possible to obtain important information about the characteristic classes of the normal bundle of a foliation.

The group most naturally related to a p-plane field is the group of linear transformations of $\mathbb{R}^m$ which take a given p-dimensional subspace onto itself. Because the natural action of GL_n on $\mathbb{R}^n$ admits no proper invariant subspaces, the p-dimensional subspace is uniquely determined by the group. Moreover, any two such subgroups are conjugate by any linear transformation which takes the one invariant subspace onto the other. The group which leaves invariant a p-dimensional subspace admits a deformation retraction onto the group of linear transformations which also leave invariant a $q = m - p$ dimensional complement. Thus, the existence of a p-plane field is equivalent to the existence of a pair of complementary p- and q-plane fields. On the other hand, the integrability problem for the pair is quite a different matter from that for a plane field. The plane field also has associated with it a principal GL_p bundle and a quotient principal GL_q bundle. Given a p-dimensional foliation without an integrable complement, it is more natural to consider these bundles than the bundle associated with a complementary pair, but the Lie algebra for the latter group admits an Ad-invariant complement in $\mathfrak{gl}_m$, while that for the group leaving invariant a single subspace does not. Moreover, a $GL_{p,q;m}$ structure has associated with it a pair of projection tensors which induce decompositions of all the tensor spaces on the manifold. Thus, the general theory of both kinds of structures will be developed, including the theory of connections and torsion. Note that the importance of the quotient bundle may be seen by looking at the structure of a Cartesian product $M \times N$, where the projections mapping onto M and N are naturally defined, but any injection of M or N into $M \times N$ requires the choice of a point in the other factor. Given a foliation, there are certain particularly interesting connections in the quotient bundle having some of their curvature forms 0, so that some of their characteristic classes vanish. The vanishing of the curvature forms, however, is a stronger condition which makes it possible to define new sets of cohomology classes, the so-called secondary characteristic classes. These characteristic classes are related to the classifying spaces for topological groupoids which will be discussed in the next chapter.

It is also interesting to consider further reductions of the foliated structure, but this consideration will be deferred to later chapters.

5.1 Definition. Let $\mathbb{R}^p$ be embedded in $\mathbb{R}^m$ as the set of vectors with the last $q = m - p$ components 0, and let $\mathbb{R}^q$ be embedded as the set of vectors with the first p components 0. Then $GL_{p;m}$ is the subgroup of GL_m that maps $\mathbb{R}^p$ onto itself and $GL_{p,q;m}$ is the subgroup of GL_m that takes both $\mathbb{R}^p$ and $\mathbb{R}^q$ onto themselves. The Lie algebras of these groups are denoted by $\mathfrak{gl}_{p;m}$ and $\mathfrak{gl}_{p,q;m}$ respectively.

In accord with the notation of I.4, $\mathbb{R}^q \times \mathbb{R}^q$ will have coordinates $(x^1, \ldots, x^p, y^1, \ldots, y^q)$. Furthermore, the indices a, b, c will run from 1 to p and the indices α, β, γ from 1 to q. It is convenient to consider bases $\{e_i\}$ for $\mathbf{R}^m$ so that $e_a = e'_a \in \mathbf{R}^p$ and $e_{p+\alpha} = e''_\alpha \in \mathbb{R}^q$. A vector is written with respect to such a basis as

$$v = v^i e_i = v^a e'_a + v^\alpha e''_\alpha$$

with the analogous notation for other tensors. In considering $GL_{p;m}$ structures, it is important to keep in mind that the second factor $\mathbb{R}^q$ is not natural, so that this notation, while convenient, is somewhat deceptive.

5.2 Proposition. *$GL_{p,q;m}$ is isomorphic to the direct sum $GL_p \oplus GL_q$, while the quotient of $GL_{p;m}$ by $GL_{p,q;m}$ is topologically equivalent to $\mathbb{R}^{pq}$. There is an epimorphism from $GL_{p;m}$ onto GL_q. The components of elements of $E^k GL_{p;m}$ and $\mathfrak{E}^k \mathfrak{gl}_{p;m}$ satisfy*

$$M^{p+\alpha}_{i_1 \ldots i_r} = 0, \qquad r \leqq k$$

provided some $i_\mu \leqq p$. The components of elements of $E^k GL_{p,q;m}$ and $\mathfrak{E}^k \mathfrak{gl}_{p,q;m}$ satisfy these conditions and also

$$M^{a}_{i_1 \ldots i_r} = 0, \qquad r \leqq k$$

provided some $i_\mu > p$. $\mathfrak{E}^k \mathfrak{gl}_{p,q;m}$ has as $Ad(E^k GL_{p,q;m})$-invariant complement in $\mathfrak{E}^k \mathfrak{gl}_{p;m}$ and in $\mathfrak{gp}^k_m$, but $\mathfrak{gl}_{p;m}$ has no $Ad(GL_{p;m})$-invariant complement in $\mathfrak{gl}_m$.

Proof. The direct sum decomposition arises by restricting each element of $GL_{p,q;m}$ to $\mathbb{R}^p$ and to $\mathbb{R}^q$. The epimorphism from $GL_{p;m}$ onto GL_q is induced by the projection of $\mathbb{R}^m$ onto $\mathbb{R}^q$ with kernel $\mathbb{R}^p$. Written with respect to a product basis, an element of $GL_{p;m}$ has the form

$$M(e'_a) = M^b_a e'_b$$
$$M(e''_\alpha) = M^a_\alpha e'_a + M^\beta_\alpha e''_\beta,$$

where M^b_a and M^β_α are nonsingular. Conversely, any collection of coefficients satisfying the nonsingularity conditions defines an element of $GL_{p;m}$. Thus, the quotient of $GL_{p;m}$ by $GL_{p,q;m}$ may be identified with $\mathbb{R}^{pq}$ with coordinates M^a_α, $GL_{p;m}$ is defined by the condition $M^\alpha_a = 0$, and $GL_{p,q;m}$ is defined by the additional condition that $M^a_\alpha = 0$. Also, the components of $E^k GL_{p;m}$ or $\mathfrak{E}^k \mathfrak{gl}_{p;m}$ must satisfy

$$M^{p+\alpha}_{ai_2 \dots i_r} = 0, \qquad r \leqq k$$

which together with symmetry of the lower indices implies the condition of the theorem. A similar calculation gives the additional condition for $E^k GL_{p,q;m}$ and $\mathfrak{E}^k \mathfrak{gl}_{p,q;m}$. In other words, the nonzero components for any element of an extension of $GL_{p,q;m}$ must be of the form $M^b_{a_1 \dots a_r}$ or $M^\beta_{\alpha_1 \dots \alpha_s}$. For $GL_{p;m}$ there are additional components of the form

$$M^a_{i_1 \dots i_r}$$

where at least one $i_\mu > p$. To show that these form an $Ad(E^k GL_{p,q;m})$-invariant complement in $\mathfrak{E}^k \mathfrak{gl}_{p;m}$, it is sufficient to show that this complement is invariant under $ad(\mathfrak{E}^k \mathfrak{gl}_{p,q;m})$ and under the adjoint action of some orientation reversing element of $GL_{p,q;m}$. In the notation of formal vector fields, elements of $\mathfrak{E}^k \mathfrak{gl}_{p,q;m}$ have the form

$$\sum_{r=1}^{k} \{M^b_{a_1 \dots a_r} x^{a_1} \dots x^{a_r} \partial/\partial x^b + M^\beta_{\alpha_1 \dots \alpha_r} y^{\alpha_1} \dots y^{\alpha_r} \partial/\partial y^\beta\}$$

while the complement consists of elements of the form

$$\sum_{\substack{r+s=1 \\ s \geqq 1}}^{k} M_{a_1 \dots a_r \alpha_1 \dots \alpha_s} x^{a_1} \dots x^{a_r} y^{\alpha_1} \dots y^{\alpha_s} \partial/\partial x^a.$$

Clearly, the bracket of an element of $\mathfrak{E}^k \mathfrak{gl}_{p,q;m}$ with an element of the complement belongs to the complement, as required. Also, the adjoint action of a linear transformation which differs from the identity by a change of sign of one coefficient merely changes the sign of some components of the vector field, so also preserves the complement. The existence of an invariant complement in $\mathfrak{gp}^k_m$ is proved similarly.

Suppose $\mathfrak{gl}_{p;m}$ admitted an $Ad(GL_{p;m})$-invariant complement $\mathfrak{m}$ in $\mathfrak{gl}_m$, and consider the adjoint operation in $\mathfrak{m}$ of an element B of $GL_{p;m}$ such that $B^b_a = \delta^b_a$ and $B^\beta_\alpha = \delta^\beta_\alpha$. Then for every $M \in \mathfrak{m}$ and every such $B \in GL_{p;m}$, there is an $N \in \mathfrak{m}$ such that

$$\begin{cases} N^\beta_a = M^\beta_a \\ N^\beta_\alpha = B^c_\alpha M^\beta_c + M^\beta_\alpha. \end{cases}$$

These equations imply that for all B, $B^c_\alpha M^\beta_c = 0$, so M lies in $\mathfrak{gl}_{p;\,m}$, a contradiction. This completes the proof of Proposition 5.2. ▯

5.3 Corollary. *If a manifold M has a $GL_{p,\,q;\,m}$ structure then every connection in $J^1 M$ induces a connection in the $GL_{p,\,q;\,m}$-structure, but the corresponding statement for $GL_{p;\,m}$ structures is false. On the other hand, a connection in a $GL_{p;\,m}$ structure induces one in any underlying $GL_{p,\,q;\,m}$ structure, and also in the quotient GL_q-bundle. A connection in a $GL_{p,\,q;\,m}$ structure induces one in the quotient GL_p and GL_q bundles. The group of any $GL_{p;\,m}$ bundle can be reduced to $GL_{p,\,q;\,m}$.*

Proof. Most of these statements follow immediately from standard results (Kobayashi and Nomizu [1963, pp. 79–83]). The last statement follows from the fact that the quotient of $GL_{p;\,m}$ by $GL_{p,\,q;\,m}$ is contractible. ▯

Though the preceding corollary is stated for subbundles of $J^1 M$, many of the statements hold for any principal bundle involving the groups under discussion. However, the special properties of $J^1 M$ are required in generalizing the definition of the tautological form θ to a form with values in $\mathbb{R}^q$ on the quotient bundle associated to a plane field of codimension q. Recall that if $x \in J^1 M$, $X \in T_x J^1 M$, and $\pi\colon J^1 M \to M$, then

$$\theta(X) = x^{-1}(T\pi X)$$

where x is viewed as an isomorphism from $\mathbb{R}^m$ to $T_{\pi x} M$ (see Kobayashi and Nomizu [1963, pp. 55–56]). If x is a point of the $GL_{p;\,m}$-structure belonging to a p-plane field, then it maps the first p vectors into the plane field, so induces an isomorphism of $\mathbb{R}^m/\mathbb{R}^p$ onto a fiber of the quotient bundle of TM by the plane field. This map depends only upon the projection ${}_q x$ of x into the quotient GL_q bundle of the $GL_{p;\,m}$ structure. Let ${}_q\pi$ be the projection of TM onto this quotient bundle, and regard ${}_q x$ as an isomorphism from $\mathbb{R}^q$ to a fiber of this bundle.

5.4 Definition. ${}_q\theta$ is the form on the quotient GL_q bundle of a plane field of codimension q defined by

$${}_q\theta(X) = {}_q x^{-1}({}_q\pi(T\pi X))$$

where X is tangent to the GL_q-bundle at ${}_q x$ and ${}_q\pi$ is the projection in this bundle. Given any connection in the quotient bundle with exterior covariant differentiation D, the *torsion* is the $\mathbb{R}^q$-valued horizontal 2-form

$${}_q\Theta = D\,{}_q\theta.$$

Since the concept of torsion has already been defined for various other bundles related to the plane field, it is important to clarify the relations among these concepts. Suppose that

$$J^1 M \supseteq E(GL_{p;\,m}, M) \supseteq E(GL_{p,\,q;\,m}, M)$$

are structures on M, the last two having the quotient bundle $E(GL_q, M)$. Let π_q denote the projection of $GL_{p;\,m}$ or $GL_{p,\,q;\,m}$ onto GL_q, and also the corresponding projections in the Lie algebras and in principal bundles.

5.5 Proposition. *The restriction of θ to $E(GL_{p,q;m}, M)$ is the sum of a form θ' with values in $\mathbb{R}^p$ and a form θ'' with values in $\mathbb{R}^q$, and the torsion with respect to any $GL_{p,q;m}$ connection has a decomposition $\Theta = \Theta' + \Theta''$ where $\Theta' = D\theta'$ and $\Theta'' = D\theta''$. If $E(GL_q, M)$ has a connection projected from $E(GL_{p,q;m}, M)$ or $E(GL_{p;m}, M)$, then $\theta'' = (T^*\pi_q)({}_q\theta)$ and $\Theta'' = (T^*\pi_q)({}_q\Theta)$.*

Proof. The decomposition $\theta = \theta' + \theta''$ follows immediately from the fact that $x \in E(GL_{p,q;m}, M)$ considered as a mapping $\mathbb{R}^p \oplus \mathbb{R}^q$ into $T_{\pi x} M$ respects the given direct sum splittings. By definition

$$\Theta(X, Y) = d\theta(HX, HY)$$

so it suffices to evaluate $d\theta$ on the standard horizontal vector fields. For these fields

$$d\theta(X, Y) = -\theta([X, Y])$$

so

$$\begin{aligned}\Theta(X, Y) &= -\theta'([X, Y]) - \theta''([X, Y]) \\ &= \Theta'(X, Y) + \Theta''(X, Y).\end{aligned}$$

Let ${}_q\theta$ be the tautological form on the quotient bundle. Then $(T^*\pi_q)\,{}_q\theta(X) = {}_q\theta(T\pi_q X) = 0$ if and only if the projection of $T\pi_q X$ into TM is tangent to the p-plane field, that is, the projection of X into TM is tangent to the p-plane field. $\theta''(X) = 0$ under the same conditions, so the two forms are equal. Also, for horizontal fields X and Y

$$\begin{aligned}\Theta''(X, Y) &= d\theta''(X, Y) \\ &= (T^*\pi_q d_q\theta)(X, Y) \\ &= d_q\theta(T\pi_q X, T\pi_q Y) \\ &= {}_q\Theta(T\pi_q X, T\pi_q Y) \\ &= (T^*\pi_q {}_q\Theta)(X, Y).\end{aligned}$$

This concludes the proof of Proposition 5.5. □

5.6 Corollary. *If ω is a connection in $J^1 M$ with $\Theta = 0$, then the induced connection in $E(GL_{p,q;m}, M)$ satisfies $\Theta' = 0$ and $\Theta'' = 0$, while the connection projected into $E(GL_q, M)$ satisfies ${}_q\Theta = 0$.*

Proof. Clearly $\Theta' = 0$, $\Theta'' = 0$, and $T^*\pi_q({}_q\Theta) = 0$. Since each horizontal space in $E(GL_q, M)$ is the image by $T\pi_q$ of a horizontal space, ${}_q\Theta = 0$. □

5.7 Corollary. *$E(GL_{p,q;m}, M)$ and $E(GL_q, M)$ have torsionfree connections.*

Proof. $J^1 M$ has such a connection. □

5.8 Remark. In the theory of foliations and plane fields, it is sometimes desirable not to choose a torsionfree connection, but to give up this property in order to obtain others that are more useful.

5.9 Lemma. *Let $E(GL_q, M)$ be the quotient bundle of a p-plane field or a pair of complementary plane fields, $x \in E(GL_q, M)$, π the projection in this quotient bundle, and ${}_q\Theta$ its torsion. Furthermore, let $X, Y \in T_{\pi x} M$, $T\pi X^* = X$, and $T\pi Y^* = Y$. Then ${}_qT(X, Y) = x_q \Theta(X^*, Y^*)$ is an element of the quotient bundle of TM by the p-plane field which depends only on X, Y, and πx. If ${}_qT$ is regarded as having values in a fixed complementary plane field, then it is a tensor such that ${}_qT(X, Y) = -{}_qT(Y, X)$, so is a vector-valued 2-form.*

Proof. It is sufficient to consider X^* and Y^* horizontal. Any two choices of x with the same projection are related by the operation of an element of GL_q which also relates the corresponding X^* and Y^*. From this the result follows easily. ▯

5.10 Definition. ${}_qT$ is called the *torsion tensor* of the connection, though this usage is imprecise in that the definition depends upon the choice of a complementary plane field. In particular, if $p = 0$, ${}_qT$ is naturally a tensor which depends only on the torsion Θ in $J^1 M$, and which will therefore be denoted by T.

The technical advantage of a $GL_{p,q;m}$ structure over a $GL_{p;m}$ structure arises from the existence of linear projections in each tangent space that are naturally associated to the structure. These projections lead to the decomposition of any tensor into pieces which frequently reflect important properties of the structure. This idea has already been used in the proof of Proposition 5.2, but it will now be developed more explicitly. The contruction proceeds as follows:

Given a $GL_{p,q;m}$ structure on M, let P denote the projection of the tangent bundle onto the first subbundle with kernel the second subbundle, and let Q denote the complementary projection. Then any tangent vector at $x \in M$ can be written uniquely in the form $X = PX + QX$, and the value of any exterior form ω of degree t at x can be written

$$\omega(X_1, \ldots, X_t) = \omega(PX_1 + QX_1, \ldots, PX_t + QX_t)_x = \sum_{r+s=t} (\pi_{r,s}\omega)(X_1, \ldots, X_t)$$

where

$$(\pi_{r,s}\omega)(X_1, \ldots, X_t) = \sum_{\pi} \operatorname{sgn} \pi\, \omega(PX_{\pi(1)}, \ldots, PX_{\pi(r)}, QX_{\pi(r+1)}, \ldots, QX_{\pi(t)}),$$

and the latter sum is over all permutations π such that $\pi(1) < \ldots < \pi(r)$ and $\pi(r+1) < \ldots < \pi(t)$. For 1-forms ω, let $P\omega = \pi_{1,0}\,\omega$ and $Q\omega = \pi_{0,1}\,\omega$. Then the dual of the preceding construction defines $\pi_{r,s}$ for multivectors. Analogous constructions define decompositions of symmetric tensors and of mixed covariant and contravariant tensors.

5.11 Definition. A form ω at a point is of *type* (r, s) if $\omega = \pi_{r,s}\,\omega$. A form ω defined on an open set is of *type* (r, s) on the set if it is of type (r, s) at each point of the set. It is of *pure type* if it is of type (r, s) for some pair (r, s). The analogous terminology is used for multivectors and mixed tensors.

The projections P and Q may also be used to define a tensor intrinsically associated to the $GL_{p,q;m}$-structure, which is a tangent vector valued exterior

2-form, just as the tensors ${}_qT$ and T are, and in fact turns out to be obtainable as the torsion of a suitably chosen connection.

5.12 Definition. The *integrability tensor* H of a $GL_{p,q;m}$ structure is the vector valued exterior 2-form defined by

$$H = \tfrac{1}{2}\{P[P,P] + Q[Q,Q]\}$$

where the bracket symbol is that defined by Nijenhuis [1951] and P and Q are regarded as vector valued 1-forms.

5.13 Proposition. *If $\{X_a\}$ is a basis for the first plane field and $\{Y_\alpha\}$ is a basis for the second, then*

$$H(X_a, X_b) = Q[X_a, X_b]$$
$$H(Y_\alpha, Y_\beta) = P[Y_\alpha, Y_\beta]$$
$$H(X_a, Y_\alpha) = 0.$$

There exist connections in any $GL_{p,q;m}$ structure such that the torsion tensor is H. There is a unique derivation D_H of the tensor algebra which commutes with contraction and satisfies:

(i) *For every function f, $D_H f = df(H)$.*
(ii) *For every vector field X*

$$D_H X = -P\mathscr{L}_{PX}H - Q\mathscr{L}_{QX}H.$$

Proof. Using the standard formula for the Nijenhuis bracket (see Kobayashi and Nomizu [1963, p. 37]), one obtains

$$\tfrac{1}{2}[P,P](X_a, X_b) = \tfrac{1}{2}[Q,Q](X_a, X_b) = Q[X_a, X_b]$$
$$\tfrac{1}{2}[P,P](Y_\alpha, Y_\beta) = \tfrac{1}{2}[Q,Q](Y_\alpha, Y_\beta) = P[Y_\alpha, Y_\beta]$$
$$\tfrac{1}{2}[P,P](X_a, Y_\alpha) = \tfrac{1}{2}[Q,Q](X_a, Y_\alpha) = 0,$$

which proves the formulas. The existence of such a connection is proved by Walker [1961, p. 98]. The derivation exists provided (i) and (ii) satisfy

$$D_H(fg) = (D_H f)g + f(D_H g)$$
$$D_H(fX) = (D_H f) \otimes X + f D_H X.$$

These are easy to verify, by using the formulas for the components of H given in the statement of the proposition. This completes the proof. □

With this background, it is time to consider the effect of integrability conditions. Since the integrability of a $GL_{p;m}$ structure does not imply that there exists an underlying integrable $GL_{p,q;m}$ structure, it is necessary to consider the two integrability conditions separately. It is natural to begin with the more fundamental integrable $GL_{p;m}$ structure, or foliation. An underlying $GL_{p,q;m}$ structure, in general not integrable, will be assumed whenever convenient. The following proposition summarizes both the integrability properties of $GL_{p;m}$ structures and the more direct integrability theory for plane fields given in § I.4, thereby establishing the relation between the two approaches.

5.14 Proposition. *Let M be a manifold of class C^l with $l > k \geqq 1$. Then M has $GL_{p;m}$ structure $E(GL_{p;m}, M)$ of class C^k if and only if in the neighborhood of each point there is a C^{k-1} basis for the tangent bundle with the first p vectors belonging to the plane field defined by the structure. For such a structure, the following conditions are equivalent.*

(i) *The structure is locally C^{k-1} isomorphic to the bundle over $\mathbb{R}^m$ of frames such that the first p vectors belong to $\mathbb{R}^p \times \{0\}$ in $\mathbb{R}^m$.*

(ii) *There is a C^k atlas of coordinate neighborhoods $\{U_\mu\}$ with coordinates (x_μ^a, y_μ^α) such that the vectors $\partial/\partial x_\mu^a$ belong to the given p-plane field.*

(iii) *There is a C^k atlas of coordinate neighborhoods $\{U_\mu\}$ with coordinates (x_μ^a, y_μ^α) such that $\partial y_\mu^\alpha/\partial x_\nu^a = 0$.*

(iv) *There is a C^k atlas of coordinates neighborhoods $\{U_\mu\}$ with coordinates (x_μ^a, y_μ^α) such that the p-plane field is defined by $dy_\mu^\alpha = 0$.*

(v) *In the neighborhood of each point there is a complete system $\{T_a\}$ of C^k flows whose generators $\{X_a\}$ span the p-plane field.*

If $k \geqq 2$, then the following conditions are also equivalent to the preceding conditions.

(vi) *In the neighborhood of each point there exist C^{k-1} closed forms ω^α such that the p-plane field is defined by $\omega^\alpha = 0$.*

(vii) *Each local C^{k-1} basis $\{X_a\}$ for the p-plane field can be extended to a basis $\{X_a, N_\alpha\}$ for the tangent bundle so that the commutator of any two elements is tangent to the p-plane field.*

If these conditions hold, then the total space of the quotient GL_q-bundle has an integrable $GL_{p;m+q^2}$ structure of class C^{k-l} such that the underlying p-plane field consists of

$$\{X \mid i_X d_q\theta^\alpha = 0\}$$

where $_q\theta$ is the tautological form. Furthermore, on each U_μ in (iv), *there is a C^{k-1}-isomorphism between the quotient bundle of the foliation and the pull-back of the tangent bundle of $\mathbb{R}^q$.*

Proof. By definition, a structure $E(G, M)$ is of class C^k if locally it has sections of class C^{k-1}. However, in case $G = GL_{p;m}$, such a section is a frame with the first p vectors belonging to the p-plane field which at πx consists of the vectors xv where $v \in \mathbb{R}^p \times \{0\}$. This proves the characterization of a $GL_{p;m}$ structure in terms of bases. To study the integrability conditions, introduce the usual coordinates in $\mathbb{R}^m$. Then the foliation whose leaves are translates of $\mathbb{R}^p \times \{0\}$ is defined by $dy^\alpha = 0$, and the vectors $\{\partial/\partial x^a\}$ form a basis for the tangent space to the leaves at any point. Since a C^k isomorphism of G-structures implies a C^k diffeomorphism on the base space, any local C^k isomorphism with the standard $GL_{p;m}$ structure can be used to define C^k coordinates (x^a, y^α). The equivalence of the first three statements is then an immediate consequence of Proposition 3.2. Their equivalence with the fourth follows because the sets $\{\partial/x_\mu^a\}$ and $\{dy_\mu^\alpha\}$ mutually define each other by the relation $dy_\mu^\alpha(\partial/\partial x_\mu^a) = 0$. Since statements (iv) through (vii) are equivalent by Proposition I.4.10^k, all seven statements are equivalent. The coordinate systems just introduced are also useful for the study of the quotient bundle, but first it is necessary to extend them to $E(GL_{p;m}, M)$. Any system of coordinates (z^i) on the base gives rise to a system (z^i, z^i_j) on $J^1 M$, and if these coordinates are defined by a local equivalence to the standard G structure, then

$$\{(z^i_j) \mid (z^i, z^i_j) \in E(G, M)\} = G.$$

In the case of $GL_{p;m}$, this means $z^{p+\alpha}_a = 0$, so that the restriction of θ to $E(GL_{p;m}, M)$ is uniquely determined by

$$dz^a = z^a_b \theta^b + z^a_\beta \theta^\beta$$
$$dz^{p+\alpha} = z^{p+\alpha}_{p+\beta} \theta^\beta.$$

Suppose the coordinates on M are (x^a, y^α) and introduce $y^\alpha_\beta = z^{p+\alpha}_{p+\beta}$. Then the quotient bundle has coordinates $(x^a, y^\alpha, y^\alpha_\beta)$ and on it

$$dy^\alpha = y^\alpha_\beta \, {}_q\theta^\beta$$

determines the tautological form. If (η^β_α) is the inverse matrix to (y^α_β), this gives

$${}_q\theta^\beta = \eta^\beta_\alpha \, dy^\alpha$$
$$d\, {}_q\theta^\beta = d\eta^\beta_\alpha \wedge dy^\alpha$$

from which it is readily seen that

$$\{X \mid i_X d\theta^\alpha = 0\} \quad \text{is generated by } \left\{\frac{\partial}{\partial X^a}\right\},$$

hence is tangent to a foliation of dimension p. Since the coordinate system on $E(GL_{p;m}, M)$ is of class C^{k-1}, this foliation is a $GL_{p;m+q^2}$-structure of class C^{k-1} on $E(GL_{p;m}, M)$. Thus, the proof of Proposition 5.14 is complete. □

5.15 Remark. In Chapter I, two different inequivalent notions of integrability for a C^k plane field were introduced. The preceding proposition shows that it is the weaker one, contained in Proposition I.4.10^k, which arises naturally in the context of G-structures. Throughout the rest of this book, the weaker definition will be used unless an explicit statement is made to the contrary.

5.16 Remark. It is now clear why H is called the integrability tensor of the $GL_{p,q;m}$ structure. Indeed, by Proposition 5.13, H vanishes if the $GL_{p,q;m}$ structure is integrable, since then both plane fields are integrable in the strong sense and a fortiori in the weak sense also. Each plane field in fact has an integrability tensor of its own, since the tensorial character of H implies that A_P defined by

$$A_P(X_a, X_b) = Q[X_a, X_b]$$
$$A_P(X_a, Y_\alpha) = A_P(Y_\alpha, Y_\beta) = 0$$

is also a tensor, the integrability tensor of the p-plane field. Similarly, the other field has an integrability tensor A_Q.

5.17 Remark. If $A_P = 0$, then the operator D_H is of type $(0,2)$, in the sense that it takes a form or multivector of type (r,s) to one of type $(r, s+2)$. Indeed, H is of type (0, 2) with respect to its arguments, so that for a function f, $D_H f$ is of type (0, 2). For a vector field X, calculation leads to $D_H X(X_a, X_b) = 0 = D_H X(X_a, Y_\alpha)$ so $D_H X$ is also of type (0, 2) with respect to its arguments. If $A_Q = 0$, then D_H is of type (2, 0), and if $H = 0$, then $D_H = 0$.

In the case of a foliation, there is a class of connections in the principal quotient bundle which is of great interest. Indeed, since the foliation of this bundle constructed in Proposition 5.14 is transverse to the fibers, it is reasonable to try to include it in the horizontal space of the connection, and to expect good things to result from that inclusion.

5.18 Definition. A connection in the quotient bundle of a $GL_{p;m}$ structure is *adapted* if the natural p-dimensional foliation of the quotient bundle is contained in the horizontal space of the connection.

Since a foliation is locally like a fiber bundle, and its normal bundle is locally like the pullback of the tangent bundle of the quotient space, it is also natural to look for connections which are locally pullbacks of connections in the tangent bundle of the quotient space. Since any pullback bundle admits pullback connections (Kobayashi and Nomizu [1963, p. 81]), the existence of such a connection is global property of the foliation.

5.19 Definition. A connection ω in the quotient bundle of a foliation is *basic* if every point has an open neighborhood with foliation coordinates (x^a, y^α) and a connection ω_U on the tangent bundle of the open set $y(U)$ in $\mathbb{R}^q$ such that $(T^* y)\,\omega_U$ is carried into the restriction of ω to U by the isomorphism of bundles.

In order to work with these definitions, it is necessary to establish in detail the relationship between connections in the quotient GL_q bundle of a foliation and covariant differentiation in the associated vector bundle, which is the quotient of the tangent bundle by the plane field of the foliation. Proofs will be omitted, since the theory parallels that for the frame bundle and tangent bundle of a manifold (see Kobayashi and Nomizu [1963, pp. 140–146]). Consider a coordinate system $(x^a, y^\alpha, y^\beta_\alpha)$ in the GL_q bundle, as in the proof of Proposition 5.14. Also, let $z^a = x^a$, $z^{p+\alpha} = y^\alpha$. Then the natural foliation is generated by the vector fields $\{\partial/\partial X^a\}$, and

$$_q\theta^\beta = \eta^\beta_\alpha\, dy^\alpha$$

where (η^β_α) is the inverse matrix to (y^β_α). Let $\{E^\alpha_\beta\}$ be a basis of $\mathfrak{gl}_q$. Then any connection form can be written

$$\omega = \omega^\beta_\alpha\, E^\alpha_\beta$$

and its pullback by the identity cross-section can be written

$$\Gamma^\beta_{j\alpha}\, dz^j\, E^\alpha_\beta.$$

Given two such coordinate systems $(x^\alpha_\mu, y^\alpha_\mu)$ and $(x^\alpha_\nu, y^\alpha_\nu)$, the $\Gamma^\beta_{j\alpha}$ are related by the formula

$$_\mu\Gamma^\beta_{j\alpha} = {}_\nu\Gamma^\delta_{i\gamma}\,\frac{\partial z^i_\nu}{\partial z^j_\mu}\frac{\partial y^\gamma_\nu}{\partial y^\alpha_\mu}\frac{\partial y^\beta_\mu}{\partial y^\delta_\nu} + \frac{\partial^2 y^\gamma_\nu}{\partial z^j_\mu\,\partial y^\alpha_\mu}\frac{\partial y^\beta_\mu}{\partial y^\gamma_\nu}$$

in which it must be remembered that

$$\frac{\partial y^\gamma_\nu}{\partial x^\alpha_\mu} = 0.$$

Conversely, given a collection of ${}_\mu\Gamma^\beta_{j\alpha}$ related by the above transformation law a connection form ω is well-defined on the principal bundle by

$$\omega^\alpha_\beta = \eta^\alpha_\gamma (dy^\gamma_\beta + \Gamma^\gamma_{j\delta}\, y^\delta_\beta\, dz^j).$$

Consider now the associated vector bundle. Any vector field Y on an open set U defines a section ϕ in this quotient bundle over U, and conversely any section ϕ can be represented by a vector field, which is not unique unless the foliation has dimension 0. There is a one-one correspondence between sections of the vector bundle and functions f from the GL_q bundle to $\mathbb{R}^q$ such that $f(xg) = g^{-1} f(x)$ for every $x \in E(GL_q, M)$ and $g \in GL_q$. This correspondence is given by

$$f(x) = x^{-1}\, \phi(\pi x).$$

If X is any vector field with horizontal lift X^*, then the covariant derivative $\nabla_X \phi$ is the section which corresponds to the function $X^* f$. Since the horizontal lift of $\partial/\partial z^j$ is

$$\frac{\partial}{\partial z^j} - \Gamma^\alpha_{j\beta}\, y^\beta_\gamma \frac{\partial}{\partial y^\alpha_\gamma}$$

and $\partial/\partial y^\alpha$ corresponds to the function $\eta^\beta_\alpha e_\beta$, where $\{e_\beta\}$ is a basis for $\mathbb{R}^q$, one has

$$\nabla_{\partial/\partial z^j}\left(\frac{\partial}{\partial y^\alpha}\right) = \Gamma^\beta_{j\alpha} \frac{\partial}{\partial x^\beta}.$$

Given a differentiation operator with suitable properties, a connection ω is defiend from it by use of the functions $\Gamma^\beta_{j\alpha}$. Suppose that for every vector field X of the manifold and every section ϕ of the quotient bundle, a section $\nabla_X \phi$ of the quotient bundle is given, so that

$$\begin{aligned} \nabla_{X+Y}\phi &= \nabla_X \phi + \nabla_Y \phi \\ \nabla_{fX}\phi &= f\nabla_X \phi \\ \nabla_X(\phi + \psi) &= \nabla_X \phi + \nabla_X \psi \\ \nabla_X(f\phi) &= (Xf)\phi + f\nabla_X \phi \end{aligned}$$

where f is any function on the manifold. Then the value of $\nabla_X \phi$ at any point depends only on X and ϕ in the neighborhood of that point, so sections over a coordinate neighborhood can be used and the same properties hold. Thus $\Gamma^\beta_{j\alpha}$ can be defined by

$$\nabla_{\partial/\partial z^j}\left(\frac{\partial}{\partial y^\alpha}\right) = \Gamma^\beta_{j\alpha} \frac{\partial}{\alpha y^\beta}$$

and they behave properly under a change of coordinates, so define a connection in the GL_q bundle, having the given operator as its covariant differentiation.

The torsion and curvature can be expressed in terms of ∇ by

$$\begin{aligned} {}_qT(X, Y) &= \nabla_X(\pi Y) - \nabla_Y(\pi X) - \pi[X, Y] \\ {}_qR(X, Y) &= [\nabla_X, \nabla_Y] - \nabla_{[X, Y]} \end{aligned}$$

where X and Y are tangent vector fields and π projects the tangent bundle into the quotient bundle. These formulas can be proved by the same methods as the corresponding formulas for the tangent bundle (Kobayashi and Nomizu [1963, pp. 133–134]).

5.20 Example. Suppose the normal bundle is trivial, and let $\{\phi_\alpha\}$ be a basis for sections. This basis gives rise to a covariant differentiation by setting $\nabla_X \phi_\alpha = 0$ for all X and all α, so that for an arbitrary $\phi = C^\beta \phi_\beta$,

$$\nabla_X \phi = (XC^\beta)\phi_\beta.$$

If the sections are given in coordinates by

$$\phi_\alpha = A_\alpha^\beta(x, y)\,\partial/\partial y^\beta$$

then

$$\Gamma_{j\alpha}^\beta = -B_\alpha^\gamma(x, y)\,\partial A_\gamma^\beta(x, y)/\partial z^j,$$

where $B_\alpha^\gamma A_\gamma^\beta = \delta_\alpha^\beta$. The sections $\phi_1, \ldots, \phi_q$ define a section of the principal bundle, that is, a G-structure where G is the group with one element. Furthermore, the restriction to this structure of the connection form vanishes, so the curvature form vanishes too.

5.21 Proposition. *Let ϕ be an aribtrary section of the quotient vector bundle, Y be any vector field on the manifold which represents ϕ, and X be any vector field tangent to the foliation. Then a connection in the quotient bundle of the foliation is adapted if and only if $\nabla_X \phi$ is the section of the quotient bundle represented by $[X, Y]$. A connection in the quotient bundle is basic if and only if it is adapted and each horizontal lift of an infinitesimal automorphism is an infinitesimal automorphism of the p-dimensional lifted foliation. The quotient bundle of every foliation of class at least C^1 admits an adapted connection, and the curvature form of such a connection vanishes whenever the class is at least C^2 and both arguments are tangent to the foliation. The curvature form of a C^2 basic connection vanishes whenever one argument is tangent to the foliation. In either case, the torsion form vanishes whenever one argument is tangent to the foliation. For any foliation of codimension q, the Pontrjagin classes of the normal bundle vanish in dimensions strictly greater than $2q$, while if it has a basic connection, they vanish in dimensions strictly greater than q.*

Proof. A connection is adapted if and only if each $\partial/\partial x^a$ is horizontal, which is equivalent to $\Gamma_{a\beta}^\alpha = 0$. The formulas

$$\nabla_{\partial/\partial x^a}\left(f^\alpha \frac{\partial}{\partial y^\alpha}\right) = \frac{\partial f^\alpha}{\partial x^a}\frac{\partial}{\partial y^\alpha} + f^\alpha \Gamma_{a\alpha}^\beta \frac{\partial}{\partial y^\beta}$$

$$\left[\frac{\partial}{\partial x^a}, f^\alpha \frac{\partial}{\partial y^\alpha}\right] = \frac{\partial f^\alpha}{\partial x^a}\frac{\partial}{\partial y^\alpha}$$

show that $\Gamma_{a\beta}^\alpha = 0$ is equivalent to the required statement in the case that $X = \partial/\partial x^a$ and ϕ is represented by $f^\alpha\,\partial/\partial y^\alpha$. However, the statement is independent of the choice of representing vector field, and tensorial in X, so the result is proved. Consider now a basic connection. Since the map

$$(x^a, y^\alpha) \to (y^\alpha)$$

induces the bundle map

$$(x^a, y^\alpha, y^\alpha_\beta) \to (y^\alpha, y^\alpha_\beta)$$

and a connection form ω on the image can be written

$$\omega^\beta_\alpha = \eta^\beta_\gamma (dy^\gamma_\alpha + \Gamma^\gamma_{\varepsilon\delta}(y)\, y^\delta_\alpha\, dy^\varepsilon),$$

its pullback has the same formula. Hence, a connection is basic if and only if $\Gamma^\gamma_{a\beta} = 0$ and $\Gamma^\gamma_{\alpha\beta}$ depends on (y^ε) alone. Thus, the horizontal lifts of the coordinate vectors are

$$\partial/\partial x^a, \qquad \frac{\partial}{\partial y^\alpha} - \Gamma^\gamma_{\alpha\beta}(y^1, \ldots, y^q)\, y^\beta_\delta \frac{\partial}{\partial y^\gamma_\delta},$$

and each $\partial/\partial x^a$ commutes with every lifted coordinate vector. Since an infinitesimal automorphism has the coordinate expression

$$g^a(x, y)\frac{\partial}{\partial x^a} + f^\alpha(y)\frac{\partial}{\partial y^\alpha},$$

its lift is therefore an infinitesimal automorphism. Conversely, if the lift of $\partial/\partial y^\alpha$ is an infinitesimal automorphism, then $\Gamma^\gamma_{\alpha\beta}$ does not depend on x, so the characterization of basic connections is proved.

To construct an adapted connection in the quotient bundle, one begins by choosing a $GL_{p,\,q;\,m}$ structure underlying the $GL_{p;\,m}$ structure and a connection ∇' in the underlying structure. Then any section ϕ in the quotient bundle is represented by a unique vector field Y which is a section of the complementary bundle to the foliation. Let X denote a vector field tangent to the leaves, and Y' denote a vector field complementary to the leaves, and define:

(i) $\nabla_X \phi$ is the section of the quotient bundle represented by $[X, Y]$.
(ii) $\nabla_{Y'} \phi$ is the section of the quotient bundle represented by $\nabla'_{Y'} Y$.
(iii) $\nabla_{X+Y'} \phi = \nabla_X \phi + \nabla_{Y'} \phi$.

The operator thus defined satisfies all the necessary conditions for a covariant differentiation in the quotient bundle, so yields a connection. To evaluate the curvature and torsion forms, it is sufficient to do so on the horizontal lifts of a coordinate basis. Since for any horizontal vector fields X and Y

$$d\omega^\beta_\alpha(X, Y) = -\,\omega^\beta_\alpha([X, Y])$$

the statements about the curvature are immediate. Since

$${}_q\theta^\varepsilon\left(\frac{\partial}{\partial x^a}\right) = 0$$

$${}_q\theta^\varepsilon\left(\frac{\partial}{\partial y^\alpha} - \Gamma^\gamma_{\alpha\beta}\, y^\beta_\delta \frac{\partial}{\partial y^\gamma_\delta}\right) = \eta^\varepsilon_\alpha,$$

$d\,{}_q\theta^\varepsilon$ can also easily be evaluated on the lifted coordinate vectors. According to Proposition 4.5 any Pontrjagin class of dimension $2k$ can be represented by a projectable form on the principal bundle whose value on a $2k$-triple of vectors is a sum of terms of the form

$$\Phi(\Omega(X_1, X_2), \ldots, \Omega(X_{2k-1}, X_{2k}))$$

where Ω is the curvature form of any connection and Φ is an invariant polynomial of degree k on $\mathfrak{gl}_q$. In evaluating the projection of Φ, it is sufficient to consider X_i which belong to one of the given pair of complementary plane fields. If the connection is adapted and $2k > 2q$, some Ω must have 2 tangent arguments, so the form is 0. If the connection is basic and $2k > q$, then some Ω must have one tangent argument, and again the form is 0. Thus, in either case, the corresponding cohomology class is also 0. This completes the proof of Proposition 5.21. □

5.22 Remark. The torsion tensor of an adapted connection and the integrability tensor of the complementary plane field are both vector valued 2-forms which vanish whenever at least one argument is tangent to the foliation. However, the integrability tensor of the complementary field has values tangent to the foliation, while the torsion tensor has values in the quotient bundle and can be defined without choosing a complementary plane field.

It is important to observe that Proposition 5.21 says much more than that certain Pontrjagin classes vanish – they vanish because certain forms representing them vanish. These forms lie in the image of the Chern-Weil homomorphism (Proposition 4.5), which is generated by differential forms C_α of degree 2α, where $\alpha = 1, \ldots, q$. If α is even, the cohomology class of C_α is a Pontrjagin class, while if α is odd, its cohomology class is 0, though C_α need not be. Thus, there exist forms h_α such that $dh_\alpha = C_\alpha$, and h_α cannot vanish if C_α does not. For a basic connection of a foliation of codimension q, certain products of the C_α are 0 as forms, however, and this leads to the possibility of constructing new closed forms. The case $q = 1$ illustrates the situation simply. Then $C_1^2 = 0$, and h_1 can be chosen so that $dh_1 = C_1$. Furhtermore

$$d(h_1 C_1) = C_1^2 = 0$$

so $h_1 C_1$ is also a closed form, of degree 3. Its construction depends upon a choice of a connection and a choice of h_1, but one might hope that at least its cohomology class depends only on the foliation. There is a construction which can be carried out directly on the foliated manifold, using only the differential structures and the foliation, but which echoes the preceding construction and can easily be proved to produce a class depending only on the foliation. For simplicity, suppose the foliation is transversally orientable, so it can be defined globally by a nonzero 1-form ω. Then $d\omega = h \wedge \omega$ and by differentiating this formula, one sees that $dh = \phi \wedge \omega = C$. The cohomology class of $h \wedge C$ can easily be shown to be independent of the choice of ω and h, and examples will be given later to show that it can be nonzero. Note that C has the property of vanishing when both arguments are tangent to the foliation, just as the curvature form does. What is needed to make a satisfactory theory covering all these ideas is a systematic way of constructing forms h_α. In particular, to be satisfactory, the theory must make it possible to compute the resulting cohomology classes, at least in the most interesting cases.

One way to construct h_α is to use the fact that given two connections with characteristic forms C_α and C'_α, there is a construction giving a form h_α such that

$$dh_\alpha = C_\alpha - C'_\alpha.$$

One of the connections is taken to be an adapted connection, and the other a connection such that the form C_α vanishes. If α is odd, it suffices to take a Riemannian connection for the second connection in the normal bundle. If the normal bundle is trivial, there is another construction giving rise to more characteristic classes. Indeed, by Example 5.20, any trivialization can be viewed as a connection, and for this connection, all the forms C_α vanish, so there are h_α for all α. Other points of view on the construction of secondary characteritic classes will be considered in Chapter III.

5.23 Definition. Let $\mathbb{R}[C_1, \ldots, C_q]$ denote the polynomial algebra over $\mathbb{R}$ generated by elements C_i of degree $2i$, and let $\mathbb{R}[C_1, \ldots, C_q]_n$ denote its quotient by the ideal consisting of elements of degree greater than n. Let $\Lambda(h_{i_1}, h_{i_2}, \ldots)$ denote the exterior algebra generated by elements h_i of degree $2i-1$. Then

$$WO_q = \Lambda(h_1, h_3, \ldots, h_l) \otimes \mathbb{R}[C_1, \ldots, C_q]_{2q}, \qquad l = 1 + 2\left[\frac{n-1}{2}\right]$$

$$W_q = \Lambda(h_1, \ldots, h_q) \otimes \mathbb{R}[C_1, \ldots, C_q]_{2q}$$

These are graded commutative algebras, which are made into cochain complexes by setting

$$dh_i = C_i \qquad dC_i = 0.$$

Note that algbras similar to these, but without the truncation, have occurred in § 4 in the calculation of the cohomology of $W(\mathfrak{g})$.

Given a subset S of $\{1, \ldots, q\}$, let I denote a finite sequence of elements of S such that $i_1 < i_2 < \ldots < i_r$ and let J denote a finite sequence of elements of $\{1, \ldots, q\}$ such that $j_1 \leqq j_2 \leqq \cdots \leqq j_s$ and $\sum j_\alpha \leqq q$. Let j_0 denote the smallest element of $J \cap S$, and i_0 denote the smallest element of I, either of them being taken to be ∞ if the set from which it is selected is empty. Let $W_{q,S}$ be the subcomplex of W_q obtained by taking as generators only those h_i with $i \in S$, so W_q is obtained by taking $S = \{1, \ldots, q\}$ and WO_q by taking $S = \{1, 3, \ldots, l\}$.

5.24 Theorem. *A basis for $H^*(W_{q,S})$ consists of the elements*

$$(h_{i_1} \wedge h_{i_2} \wedge \ldots \wedge h_{i_r}) \otimes (C_{j_1} \ldots C_{j_s})$$

where $i_o + \sum j_\alpha > q$ and $i_0 \leqq j_0$. Given any sufficiently smooth foliation of codimension q on M, there is a homomorphism from $H^*(WO_q)$ to $H^*(M)$ such that for any sufficiently smooth *mapping $f: M_1 \to M_2$ transversal to foliations of codimension q, the diagram*

$$\begin{array}{ccc} & & H^*(M_2) \\ & \nearrow & \\ H^*(WO_q) & & \\ & \searrow & \\ & & H^*(M_1) \end{array}$$

is commutative. For foliations with a trivial normal bundle, there is a homomorphism from $H^*(W_q)$ to $H^*(M)$ with analogous properties.

Proof. The calculation of $H^*(W_{q,s})$ is due to Vey, and has been published in a paper of Godbillon [1972]. The homomorphisms from $H^*(W_{q,S})$ to $H^*(M)$ are constructed by using two connections, as indicated above. However, it is necessary to show that the cohomology classes obtained on M are independent of the choice of the two connections. This has been done by Bott [1972], who considers a set of n connections on a bundle over M as a connection in a bundle over $M \times \mathbb{R}^{n-1}$. This is a direct generalization of the proof that the Chern-Weil homomorphism is independent of the choice of connection. ▯

5.25 Remark. Since this method requires slightly more smoothness than is natural to the problem, the precise statement of the smoothness condition is left to Chapter III, § 2.

5.26 Definition. The relative characteristic classes of a foliation are the image of this homomorphism from $H^*(WO_q)$ to $H^*(M)$. If the foliation has trivial normal bundle, then its absolute characteristic classes are the image of $H^*(W_q)$.

Note that the relative characteristic classes include the Pontrjagin classes C_{2i} of the normal bundle. If the normal bundle is trivial, its Pontrjagin classes vanish, so they do not occur as absolute characteristic classes. All the other classes are of dimension at least $2q+1$, and the product of any two of them vanishes. To show that these classes are not always zero, it is sufficient to consider examples.

5.27 Example. In the construction of Example I.6.6, consider the hyperbolic plane H^2. The left invariant forms on its isometry group have a basis $\{\xi_1^*, \xi_2^*, \zeta_{12}^*\}$ such that

$$\begin{aligned} d\xi_1^* &= -\xi_2^* \wedge \zeta_{12}^* \\ d\xi_2^* &= \xi_1^* \wedge \zeta_{12}^* \\ d\zeta_{12}^* &= \xi_1^* \wedge \xi_2^*. \end{aligned}$$

The group admits a foliation defined by $\omega = \xi_2^* + \zeta_{12}^*$, and $d\omega = h \wedge \omega$ where $h = \xi_1^*$. The characteristic class $h_1 C_1$ is given by

$$h \wedge dh = -\xi_1^* \wedge \xi_2^* \wedge \zeta_{12}^*.$$

H^2 is the universal covering space of any compact 2-manifold M^2 of curvature -1, and the above foliation induces one on the bundle S of unit tangent vectors of M^2 because it is invariant by the deck transformations, which are isometries. The construction of h also projects to the foliation of S, and $h \wedge dh$ represents a nontrivial element of H^3 because it vanishes nowhere.

This construction generalizes to produce many examples of nontrivial characteristic classes, but not all the needed examples can be so obtained. Other constructions will be discussed in Chapter III.

5.28 Remark. For a foliation with a basic connection, the products of C_i vanish in dimension above q instead of above $2q$. Thus $\mathbb{R}[C_1, \ldots, C_q]_{2q}$ can be replaced by $\mathbb{R}[C_1, \ldots, C_q]_q$ and the construction with connections applied to give a different collection of characteristic classes. Furthermore, the restriction of h_i to any leaf is a closed form, so defines a cohomology class on the leaf which will be seen (§ III.1) to be related to the holonomy.

The study of connections for $GL_{p;m}$ structures and $GL_{p,q;m}$ structures was begun by Walker [1955, 1958] and Willmore [1956, 1957]. Their interest was primarily in integrability conditions and the torsional derivative. Bott [1970] first noticed the implications for characteristic classes, and from this came to define and construct the secondary classes by means of connections (Bott [1972]). At the same time, three other sets of authors also introduced the foliation classes. Godbillon and Vey [1971] introduced the 3-dimensional class associated with a foliation of codimension 1, and noticed the connection to the Lie algebra of formal vector fields in one variable. Kamber and Tondeur [1972, 1973, 1974] came to the definitions through their work on flat bundles. Finally, Bernshtein and Rozenfeld [1972] came to the definitions through the theory of formal vector fields. Basic connections and their characteristic classes were introduced by Molino [1971, 1972], who also studied [1973] the geometric and topological properties of foliations with a basic connection.

Chapter III. Singular Foliations

In the second chapter, certain cohomology classes have been defined on a foliated manifold. These are obtained by using differential forms related to the characteristic classes of the normal bundle. Moreover, the cohomology classes have formal properties analogous to those of the characteristic classes of a vector bundle. Hence, it is reasonable to try to obtain them by mapping into a classifying space for foliations, that is, some sort of foliated space whose cohomology consists of the characteristic classes of its foliation, and such that all foliations are obtained (up to some reasonable equivalence) by mapping into the classifying space and pulling back its foliation. Even for vector bundles, the classifying space is not a finite dimensional smooth manifold, so it is necessary to generalize the definition to allow foliations of much more general topological spaces, and so that the generalized foliations map contravariantly under a much larger class of mappings. In particular, even for a smooth manifold some class of singularities must be allowed. In the three sections of this chapter, three types of classifying spaces and the accompanying notions of singularity will be discussed.

In the context of the general theory, a q-dimensional vector bundle can be viewed as associated to a principal bundle with group GL_q, and the classifying space classifies bundles with this group. A construction can be given which generalizes to arbitrary topological groups, and even to topological groupoids. In the latter case, the objects classified are no longer bundles, but Γ-structures, which can be viewed as germs of bundles (microbundles). The characteristic classes for foliations of condimension q are cohomology classes of $B\Gamma_q^k$, $k \geqq 2$, and each Γ_q^k-structure has a bundle associated to it, generalizing the normal bundle of a foliation, so there is a map of classifying spaces

$$Bv\colon B\Gamma_q^k \to BGL_q.$$

Since a classifying space is defined only up to homotopy type, the appropriate problem is to determine the homotopy types of $B\Gamma_q^k$ and Bv. Much is known about the solution to this problem, but much remains unknown.

Another type of classifying space can be obtained by a construction based on the algebraic theory of flat connections (§ II.4). If $\mathfrak{g}$ is the Lie algebra of the Lie group G, then for any flat G-bundle, there is a homomorphism from the dual $\mathfrak{g}^*$ to the forms on the total space of the bundle which commutes with the exterior derivative. Thus, the cohomology of $\mathfrak{g}$ is mapped into cohomology classes on the total space, possibly even on the base space. This whole process can be generalized to an important class of infinite dimensional Lie algebras, namely, those whose underlying topological vector space is a nuclear space. In particular, the classifying space Ba_q^∞ is related to foliations of codimension q.

From the point of view of integrability theory (§ I.4), a foliation of codimension q is defined locally by independent forms $\omega^1, \ldots, \omega^q$ satisfying

$$d\omega^\alpha \wedge \omega^1 \wedge \ldots \wedge \omega^q = 0, \quad \alpha = 1, \ldots, q.$$

Dropping the requirement of independence leads to the notion of Frobenius structure and the classifying space $B\mathscr{F}_q^k$, $k \geqq 2$. Much remains unknown about the degree of difference among these three notions of singular foliation.

1. The Classifying Space for a Topological Groupoid

The notion of topological groupoid has already been introduced (Definition I.1.6), as well as the topological groupoid associated with a pseudogroup of transformations (Proposition I.1.7). Here it is necessary to develop the theory of topological groupoids in more detail, in order to construct the classifying space for a topological groupoid and to apply the latter to study the existence of foliations on a given manifold and the characteristic classes of these foliations.

1.1 Definition. A *homomorphism* ϕ of topological groupoids is a continuous function which takes each identity to an identity and such that if xy is defined, $\phi(x)\phi(y)$ is defined and equal to $\phi(xy)$.

1.2 Definition. A *cocycle on X with coefficients in the topological groupoid Γ* is an open covering $\{V_\mu\}$ of X and a family of continuous maps.

$$g_{\lambda\mu}: V_\lambda \cap V_\mu \to \Gamma$$

such that $g_{\lambda\nu}(x) = g_{\lambda\mu}(x)\, g_{\mu\nu}(x)$ for every $x \in V_\lambda \cap V_\mu \cap V_\nu$. Two cocycles are *equivalent* if they can be extended to a cocycle on the union of the coverings. A *Γ-structure* is an equivalence class of cocylces. It is *numerable* if it can be represented by a cocycle on $\{V_\mu = \phi_\mu^{-1}(0,1]\}$ where $\{\phi_\mu\}$ is a countable partition of unity.

Note that the convexity problem which arises in the definition of pseudogroup structures does not occur here. In the cases of interest, this is because the groupoid consists of germs rather of mappings with a fixed domain.

1.3 Definition. If $f\colon Y \to X$ is continuous and $\{V_\mu, g_{\mu\nu}\}$ is a cocycle defining a Γ structure on X, then its *inverse image* is the Γ-structure on Y defined by $\{f^{-1}(V_\mu), g_{\mu\nu} \circ f\}$. The *$\Gamma$-structure induced on a subspace* is the inverse image by the injection map. Two numerable Γ-structures on X are *homotopic* if there is a numerable Γ-structure on $X \times I$ which induces the given structures on $X \times \{0\}$ and $X \times \{1\}$ respectively.

Since the inverse image of a numerable Γ-structure is numerable and all Γ-structures on a paracompact space are numerable, the consideration of numerable structures is sufficient for present purposes. This restriction makes it possible to give a simple direct construction of a classifying space $B\Gamma$ for Γ and the universal numerable Γ-structure which it carries.

Consider a sequence of pairs

$$(\gamma_i, t_i) \qquad i = 1, 2, \dots$$

where $t_i \in [0,1]$, $\gamma_i \in \Gamma$, all the γ_i have the same right identity, all but a finite number of the t_i vanish, and $\Sigma t_k = 1$. Two such sequences are identified if $t_i = t_i'$ and if, when they are both nonzero, $\gamma_i = \gamma_i'$. An equivalence class of such sequences is denoted by

$$[\gamma, t] = [(\gamma_1, t_1), (\gamma_2, t_2), \dots]$$

and $E\Gamma$ is the set of all such equivalence classes. Let p_i: $E\Gamma \to [0,1]$ be defined by $p_i[\gamma, t] = t_i$ and S_i: $p_i^{-1}(0,1] \to \Gamma$ be defined by $S_i[\gamma, t] = \gamma_i$. $E\Gamma$ is given the coarsest topology so that the maps p_i and S_i are continuous. Γ acts on the right of $E\Gamma$ by $[\gamma, t]\beta = [\gamma\beta, t]$ provided the left identity of β is the right identity of each γ_i.

1.4 Definition. The *classifying space* $B\Gamma$ for Γ is the quotient of $E\Gamma$ by Γ, with quotient map π: $E\Gamma \to B\Gamma$. The *universal Γ-structure* on $B\Gamma$, is given by the cocycle

$$\{V_\lambda = \pi(P_\lambda^{-1}(0,1]),\ g_{\lambda\mu} = \gamma_\lambda \gamma_\mu^{-1}\}.$$

1.5 Theorem. *There is a bijection between homotopy classes of numerable Γ-structures on X and the set of homotopy classes of continuous maps of X into $B\Gamma$, where Γ is a connected topological groupoid.*

Proof. Given a cocycle $\{V_\lambda, g_{\lambda\mu}\}$ on X with associated partition of unity $\{\phi_\lambda\}$, define a function f_λ: $V_\lambda \to E\Gamma$ by

$$f_\lambda(x) = [(g_{1\lambda}(x), \phi_1(x)), (g_{2\lambda}(x), \phi_2(x)), \dots].$$

f_λ is continuous and passes to the quotient, giving to a mapping of X into $B\Gamma$. The proof that this construction has the required properties, as well as various other proofs required in the preceding constructions, can be found essentially in the paper of Milnor [1956] dealing with classifying spaces for topological groups. The definitions given above, extending his construction to topological groupoids, are due to Buffet and Lor [1970]. Their construction is the same as Milnor's in case Γ is a group, since then there is only one identity element. □

It is clear from the construction that any homomorphism h: $\Gamma \to \Gamma'$ of topological groupoids gives rise to a continuous map Bh: $B\Gamma \to B\Gamma'$. In the case of the topological groupoid $\Gamma_m^k = \Gamma(\mathscr{G}_m^k)$, $k \geqq 1$, the determinant map is a homomorphism into GL_m, which therefore gives rise to a map of $B\Gamma_m^k$ into BGL_m. In case the Γ_m^k-structure is a C^k structure on an m-dimensional manifold X, the composition $X \to B\Gamma_m^k \to BGL_m$ is the classifying map for the tangent bundle. More generally, if a Γ_q^k-structure arises from the local projections of a foliation of codimension q, this composition classifies the normal bundle of the foliation. Thus, the following definition is reasonable.

1.6 Definition. The *normal bundle* of a Γ_q^k-structure given by the cocycle $\{V_\lambda, \gamma_{\lambda\mu}\}$ is the GL_m bundle given by $\{V_\lambda, \det \gamma_{\lambda\mu}\}$.

Consider again the case that a Γ_q^k-structure arises from a foliation of codimension q. Then there is a Γ_q^k-structure induced on any leaf by its injection map. This structure is what has previously been called the holonomy of the leaf.

1.7 Definition. Let α be a singular cohomology class on $B\Gamma$ and let $f: X \to B\Gamma$ be continuous. Then $f^*\alpha \in H^*(X)$ is a *characteristic class* for the homotopy class of Γ-structures defined by f.

1.8 Remark. Suppose A is a functor which assigns to each homotopy class of Γ-structures on X an element of the singular cohomology $H^*(X)$. Suppose furthermore that if $f: Y \to X$ is continuous and ξ is any Γ-structure on X, then $A(f^{-1}\xi) = f^*(A\xi)$. Then $A(\xi)$ is a characteristic class defined by $\alpha \in H^*(B\Gamma)$, where α is the value of A on the universal Γ-structure. In fact, if A is defined for a sufficiently large class of Γ-structures, α can sometimes be found even though A does not apply directly to the universal Γ-structure. Examples of this will be seen later.

Since Γ_q^k-structures are used to study foliations, it is important to ask when a homotopy class of such structures contains a foliation, and if it does, how nearly is the foliation determined by the Γ_q^k-structure. Though any Γ_q^k-structure has an abstractly defined normal bundle, in the case of a foliation this normal bundle can be embedded in the tangent bundle of the manifold, for example as the normal bundle with respect to some Riemannian metric. This condition is sufficient for the existence of a foliation. In order to state the classification theorem, another definition is required.

1.9 Definition. Two foliations of codimension q on M are *concordant* if there is a foliation of codimension q on $M \times I$ which induces the given foliations on $M \times \{0\}$ and $M \times \{1\}$. Two monomorphisms of a given bundle v into the tangent bundle $T(M)$ are *concordant* if there is a monomorphism of v into $T(M \times I)$ which induces the given monomorphisms on $M \times \{0\}$ and $M \times \{1\}$.

1.10 Theorem. *Let M be a manifold (without boundary), let $q \geqq 1$, and let $k \geqq 1$. Then the concordance classes of C^k codimension q foliations on M correspond bijectively to pairs composed of a homotopy class of Γ_q^k-structures on M and a concordance class of bundle monomorphisms of the normal bundle into $T(M)$. If $q > 1$ and the data are consistent, the foliation can be taken to agree with some given foliation on the neighborhood of a closed subset.*

If $q = 1$, the Reeb-stability theorem prevents the last statement from being true, but precise conditions can be given for the existence of a foliation tangent to the boundary if the boundary is nonempty.

Proof. This theorem is due to Thurston [1974a, 1976]. □

It is frequently convenient to replace the notion of Γ-structure on a manifold by the equivalent notion of foliated microbundle over the manifold. In the definition, $\mathscr{G}$ is a transitive pseudogroup on $\mathbb{R}^q$ with Γ its associated topological groupoid.

1.11 Definition. A *foliated microbundle* over the manifold M consists of a manifold E, a submersion $p: E \to M$, a section $i: M \to E$ of p, and a foliation transverse to the fibers whose local coordinate changes belong to $\mathscr{G}$. Two such foliated microbundles are *equivalent* if they agree on a neighborhood of $i(M)$.

The inverse image by i of the foliation is a Γ-structure on M, and in fact this construction gives a bijection between $\mathscr{G}$-foliated microbundles and Γ-structures.

Because of the usefulness of $B\Gamma_q^k$ in classifying foliations, it is necessary to learn as much as possible about the homotopy and cohomology of this space and of the determinant map v: $B\Gamma_q^k \to BGL_q$. In fact, the case of topological Γ_q-structures ($k = 0$) is quite different from that of structures of class C^k with $k \geqq 2$. Moreover, there remain many unanswered questions about v and $B\Gamma_q^k$, and even more about other topological groupoids. In studying the map v, it is convenient to replace it by a homotopically equivalent fibration

$$B\bar{\Gamma}_q^k \to B\Gamma_q^k \to BGL_q.$$

$B\bar{\Gamma}_q^k$ is the classifying space for Γ_q^k-structures with trivial normal bundle.

1.12 Proposition. *The induced cohomology map*

$$H^r v\colon H^r(BGL_q; A) \to H^r(B\Gamma_q^k; A)$$

is 0 *for* $r > 2q$, $k \geqq 2$, *and* $A = \mathbb{R}$. *For* $A = \mathbb{Z}$ *or* $A = \mathbb{Z}/p\mathbb{Z}$ *for any prime* p, $k \geqq 1$, *and every* r, $H^r v$ *is injective, so that* $H_{4r-1}(B\Gamma_q^k; \mathbb{Z})$ *is not finitely generated for* $k \geqq 2$ *and* $2r > q$. *There is a map* w: $H^*(WO_q) \to H^*(B\Gamma_q^k; \mathbb{R})$ *for* $k \geqq 2$ *such that for any* C^k *foliation on* M, *classified by* f, $(H^*f) \circ w$ *is the natural map from* $H^*(WO_q)$ *to* $H^*(M)$ *defined in Theorem* II.5.24. $\pi_r(B\bar{\Gamma}_q) = 0$ *for* $r \leqq p + 1$, *while* $B\bar{\Gamma}_q^0$ *is contractible.*

Proof. To specify a singular cohomology class of $B\Gamma_q^k$, it is sufficient to give its value on every singular homology class. Any singular homology class may be represented by a map of a finite complex, and any homology between two classes by a map of a larger complex containing the two given complexes. Any finite complex can be embedded in euclidean space, hence has the homotopy type of a C^∞ manifold, namely any of its regular neighborhoods. Any map of a C^∞ manifold M into $B\Gamma_q^k$ gives rise to a C^k foliated microbundle E over M, to which the constructions of § II.5 can be applied. Using these ideas, Bott [1972] shows that the cohomology with real coefficients has the properties listed above. Bott and Heitsch [1972] prove the statements about cohomology with coefficients $\mathbb{Z}$ and $\mathbb{Z}/p\mathbb{Z}$. The main idea is that for any finite subgroup A of O_q, the map of the classifying space BA into BGL_q factors through a map from BA into $B\Gamma_q^k$ for

$k \geqq 1$. The statements on π_r are due to Thurston [1974c], who relates $B\bar{\Gamma}_q^k$ to the group consisting of C^k diffeomorphisms of $\mathbb{R}^q$ with compact support. Indeed, the natural map of this group with the discrete topology onto itself with the C^k topology has as homotopy fiber a group whose classifying space has the same homology groups as the q-th loop space of $B\bar{\Gamma}_q^k$. This completes the outline of the proof. □

The homomorphism w mentioned in the statement of Proposition 1.12 can provide much information about the cohomology of $B\Gamma_q^k$ if its kernel is not too big. Given a class α in $H^*(WOq)$, to show that $w^*(\alpha) \neq 0$, its suffices to find a topological space X and homomorphisms ϕ and ψ such that $\phi(\alpha) \neq 0$ and the following diagram is commutative:

$$\begin{array}{ccc} H^*(WO_q) & \xrightarrow{\;w\;} & H^*(B\Gamma_q^k; \mathbb{R}) \\ & {}_{\phi}\searrow \quad \swarrow_{\psi} & \\ & H^*(X) & \end{array}$$

One possibility is that X is a smooth manifold having a C^k foliation of codimension q and ψ is induced by the classifying map, but there are other choices possible in this construction. Furthermore, the analogous question arises for other topological groupoids, and for the cohomology of $B\bar{\Gamma}_q^k$, which is related to $H^*(W_q)$ by Theorem II.5.24.

One way to get useful examples is by a systematic study of bundles with a foliation transverse to the fibers.

1.13 Definition. Let K be a subgroup of G, and let G^δ be G with the discrete topology. A bundle with group K is said to be *G-foliated* if there is given a G-bundle equivalence with a bundle with group G^δ.

There is a classifying space $B(G^\delta, K)$ for G-foliated K-bundles. In fact

$$B(G^\delta, K) = BG^\delta \times_{BG} BK$$

where $BG^\delta \to BG$ and $BK \to BG$ are induced by group homomorphisms and $B(G^\delta, K)$ is the subset of $BG^\delta \times BK$ consisting of pairs of points with the same image in BG. The pull-back by a continuous map and characteristic classes are defined as usual. If K and G are Lie groups and the structure is C^∞, a general method for calculating the characteristic classes is given by Kamber and Tondeur ([1975] and references therein). Though restricted in this way, the situation they consider is more general in that the transverse foliation may have dimension less than that of the base manifold, provided it projects into a foliation of the base. (This is their meaning for the phrase "foliated bundle".) If $\mathfrak{g}$ is the Lie algebra of G, the characteristic classes they calculate arise from the cohomology of a truncated relative Weil algebra $W(\mathfrak{g}, K)_s$, where s depends on the codimension and the groups.

Another way to get information about the cohomology of $B\Gamma_q^k$ is by studying its rational homotopy by means of the theory of minimal models, due to Sullivan

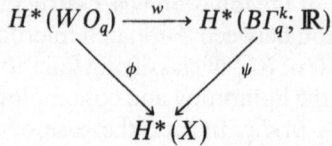

[1977]. Let $I(G)_s$ be the algebra of invariant polynomials on $\mathfrak{g}$, truncated appropriately. It is a differential graded algebra, thus has a minimal model which determines the dual rational homotopy. A connection gives rise to a homomorphism from $I(G)_s$ to the forms on the base M, and two such maps for different connections are algebra homotopic. Thus, there is a map from the dual homotopy of $I(G)_s$ to that of M and to that of the appropriate classifying space. This approach to the topology of $B\Gamma$ is due to Hurder [1980b].

A third method for proving the nontriviality of elements of $B\Gamma$ is due to Fuks [1977], who uses the cohomology of the Lie algebra of vector fields on a manifold rather than the cohomology of the manifold itself. This approach will be discussed in § 2. One of his results is that the homomorphism

$$w\colon H^*(W_q) \to H^*(B\bar{\Gamma}_q^k; \mathbb{R})$$

is injective.

Another interesting question is how the characteristic classes vary if a Γ_q^k-structure varies slightly. In the case of a homotopy, that is, a Γ_q^k-structure on $X \times I$, the classes induced on each $X \times \{t\}$ are the same. However, it may be that Γ_q^k-structures are given on each $X \times \{t\}$, varying continuously but not forming a Γ_q^k-structure on $X \times I$. Formulating this idea carefully leads to the following definition.

1.14 Definition. Let $p\colon W \to M$ be a C^k locally trivial fiber space, where the dimension of M is m. Suppose W admits a Γ_{q+m}^k-structure with coordinate transformations which project into coordinate transformations on M. Then p is called a *deformation* of Γ_q^k-structures.

Note that there is a Γ_q^k-structure induced on $p^{-1}(x)$ for each x in M. The behaviour of the characteristic classes is determined by the truncation homomorphism

$$\varrho\colon WO_{q+m} \to WO_q.$$

1.15 Proposition. *The classes in the image of*

$$\varrho^*\colon H^*(WO_{q+m}) \to H^*(WO_q)$$

are unchanged by a deformation of Γ_q^k-structures.

Proof. For $m = 1$, this theorem is due to Heitsch [1973]. A more general result than this has been proved by Kamber and Tondeur [1975, p. 85]. Heitsch [1978] has also given examples to show that those classes which are not in the image can vary independently. For the Godbillon-Vey class this was first proved by Thurston [1972]. An important tool used by Heitsch is a residue theorem for a vector field preserving a foliation. This theorem is inspired by the work of Baum and Bott [1972] in the complex case and the work of Chern and Simons [1974] on differential characters. □

Characteristic classes related to the holonomy of a leaf L and lying in $H^*(L)$ have already been defined (II.5.28). These can be related to classifying space theory by the use of G foliated K-bundles. Indeed, if K is a maximal compact subgroup of the Lie group G, then $B(G^\delta, K)$ has the homotopy type of BG^δ, and there is a homotopy fibration

$$B(G^\delta, e) \to BG^\delta \to BG.$$

The characteristic classes arising from $H^*(\mathfrak{g}, K)$ thus belong to $H^*(BG^\delta)$ and those arising from $H^*(\mathfrak{g})$ belong to $H^*(B(G^\delta, e))$. More information can also be obtained about where these classes lie in $H^*(L)$. Indeed, the map from L to BG^δ which classifies the flat structure of the normal bundle lifts to a map from the universal cover $\tilde{L}$ to the total space EG^δ. Letting $\pi = \pi_1(L)$ operate on EG^δ by means of the holonomy homomorphism into G^δ gives rise to a mapping

$$\tilde{L} \times_\pi E\pi \to EG^\delta \times_\pi E\pi$$

whose induced cohomology map identifies $H^*(B\pi)$ with one factor of the tensor product $H^*(\tilde{L}) \otimes H^*(B\pi)$ which constitutes the E_2 term in the spectral sequence for the universal cover $\tilde{L}$.

To extend these results to Γ_q-structures, it is necessary to introduce the ideas of leaf and of holonomy for such structures.

1.16 Definition. The *leaf topology* on a space X with a Γ_q-structure is the topology having as a basis the intersections of open sets of X with the level sets of the local maps into $\mathbb{R}^q$ defining the structure. A leaf is a connected component with respect to the leaf topology. The holonomy of a leaf is the restriction of the structure to the leaf.

As in the case of a foliation, the holonomy of a leaf L of a Γ_q-structure is determined by a homomorphism of the fundamental group into Γ_q, at least if the topology of the leaf is such that the fundamental group is well-defined. Thus, there are homomorphisms

$$\pi_1(L) \to \Gamma_q \to GL_q^\delta \to GL_q$$

and it is natural to ask whether there are Lie algebras whose cohomology is contained in that of $B\Gamma_q$ or $B\bar{\Gamma}_q$ and has an appropriate relation to the cohomology of $\mathfrak{gl}_q$ and the other Lie algebras already mentioned. In fact, the required Lie algebras are infinite dimensional, and the next section will be devoted to them.

2. Vector Fields and the Cohomology of Lie Algebras

In the previous section, it has been seen how the topological groupoid associated to a pseudogroup of homeomorphisms can be used to get information about pseudogroup structures on a manifold, indeed, about a more general class of structures defined on any topological space. Comparison with the behavior of Lie groups suggests that further information can be gotten if a Lie algebra can be associated to the pseudogroup. For the pseudogroup of all C^∞ diffeomorphisms on a given manifold, the Lie algebra of all C^∞ vector fields is a natural object to consider, since any vector field can be integrated locally to give a family of elements of the pseudogroup. $\mathfrak{a}_m^\infty$ is also a natural object to associate with this pseudogroup, since any vector field gives rise to an element of $\mathfrak{a}_m^\infty$ at each point of the manifold. More generally, for any transitive C^∞ pseudogroup $\mathscr{G}$, it is reasonable to consider those vector fields whose local families consist of elements of $\mathscr{G}$. One important type of subalgebra consists of compactly supported vector fields. Such a subalgebra corresponds to a group, namely, a group of compactly supported diffeomorphisms. Interesting quotients are obtained by dividing out by the subalgebra of vector fields vanishing with all their derivatives on some closed subset. If the closed subset is a single point, this quotient algebra is a subalgebra of $\mathfrak{a}_m^\infty$. Since these algebras are infinite dimensional, it is necessary to specify a topology in order to work with them. This has already beend done for $\mathfrak{a}_m^\infty$ in § II.2. Some other important cases follow.

2.1 Definition. Given a C^∞ manifold M, $v(M)$ is the Lie algebra of all C^∞ vector fields on M. $v(M)$ is given the topology of uniform convergence of all derivatives on compact subsets. $v_c(M)$ is the Lie algebra of compactly supported vector fields, and is topologized as the direct limit over compact sets of the topology of uniform convergence of all derivatives.

Topologies for infinite dimensional vector spaces will be discussed in § IV.1. For the present, a few comments will suffice. In studying these Lie algebras, it is desirable to introduce a cohomology theory which mimics as closely as possible the cohomology theory of finite dimensional Lie algebras. Thus, the topologies should be such that the set of continuous linear forms on a space and the tensor product of finitely many spaces admit topologies such that all the usual manipulations of finite dimensional linear algebra are valid. The concept of nuclear space was introduced for this purpose (Grothendieck [1951]). In general, almost the only nuclear locally convex linear spaces are finite dimensional spaces, spaces of C^∞ objects, spaces of holomorphic objects, and their duals. Since $\mathfrak{a}_m^\infty$ is nuclear, $J^\infty M$ may be viewed as an infinite dimensional manifold modelled on a nuclear space. Since the bracket operations are continuous, $\mathfrak{a}_m^\infty$, $v(M)$, and $v_c(M)$ are nuclear Lie algebras. Consequently, throughout this section all infinite dimensional structures mentioned will be assumed to be nuclear, and the properties of finite dimensional structures will be carried over without comment, except of course in those cases in which they are not true for infinite dimensional nuclear structures.

It is important to note that the implicit function theorem and the local integrability theorem for vector fields are not in general true for nuclear manifolds. However, the implicit function theorem is known to be true for a large class of nuclear manifolds, and examples show how it fails in others (Hamilton [1982]). Furthermore, though the language of nuclear structures is convenient, the available references (Gelfand, Kazhdan, and Fuks [1972] and Bernshtein and Rozenfeld [1973]) do not cover all aspects of the general theory. Consequently, the currently available proofs of certain statements apply only to the special cases of greatest interest.

Differential geometers know the value of differential forms, which in the present context may be viewed as alternating $C^\infty(M)$-multilinear $C^\infty(M)$-valued functions on the space of the vector fields. Analysts know the value of continuous $\mathbb{R}$-linear $\mathbb{R}$-valued functions on a topological vector space – in the cases of interest here, these form a nuclear space related to the distributions of Schwartz [1957]. By combining these ideas, one is led to consider alternating continuous $\mathbb{R}$-multilinear functionals on a nuclear Lie algebra. (For a more detailed motivation in the case of $v(M)$, see the paper of Bott [1976].)

2.2 Definition. Given a nuclear topological Lie algebra $\mathfrak{g}$, a *q-cochain of* $\mathfrak{g}$ (with real coefficients) is an alternating q-linear (over the reals) continuous real-valued function. If ϕ is a q-cochain, $d\phi$ is defined by

$$d\phi(X_0, \ldots, X_q) \\ = \sum_{i<j}(-1)^{i+j}\phi([X_i, X_j], X_0, \ldots, X_{i-1}, X_{i+1}, \ldots, X_{j-1}, X_{j+1}, \ldots, X_q).$$

$C^q(\mathfrak{g})$ is the set of all q-cochains and $H^q(\mathfrak{g})$ is the cohomology of the cochain complex $\{C^q(\mathfrak{g}), d\}$.

Note that $d^2 = 0$ and d is a continuous homomorphism from $C^q(\mathfrak{g})$ to $C^{q+l}(\mathfrak{g})$, so $H^q(\mathfrak{g})$ is well defined. The image of d is closed in many interesting cases, so in these cases $H^q(\mathfrak{g})$ has a nuclear Hausdorff topology. Since the topology given on $v_c(M)$ is finer than that induced as a subset of $v(M)$, $C^q(v_c(M))$ contains many elements which are not restriction of elements of $C^q(v(M))$. However, $C^q(v_c(M))$ contains all the objects necessary for analysis on M (see § IV.1).

It is frequently necessary to consider cohomology with coefficients is some more complicated object, such as the space of sections of a tensor bundle. The object must be a module over $\mathfrak{g}$ – in the case of the tensor bundle, $v(M)$ acts on the sections by Lie differentiation. The definition of d is again taken from differential geometry:

$$(2.3)\quad d\phi(X_0, \ldots, X_q) = \sum(-1)^i X_i \cdot \phi(X_0, \ldots, X_{i-1}, X_{i+1}, \ldots, X_q) \\ + \sum(-1)^{i+j}\phi([X_i, X_j], X_0, \ldots, X_{i-1}, X_{i+1}, \ldots, X_{j-1}, X_{j+1}, \ldots, X_q).$$

The case of real coefficients is obtained by letting $\mathfrak{g}$ act trivially on $\mathbb{R}$ (the derivative of a constant is 0). For some purposes, it is convenient to consider d as the dual of a boundary operator

$$\partial(X_1 \wedge \ldots \wedge X_p) = \sum_{i<j}(-1)^{i+j}[X_i, X_j] \wedge X_1 \wedge \ldots \wedge X_{i-1} \wedge X_{i+1} \wedge \ldots \\ \wedge X_{j-1} \wedge X_{j+1} \wedge \ldots \wedge X_p$$

and to introduce the homology of $\mathfrak{g}$ by means of this operator. Some references on the homology and cohomology of Lie algebras are Koszul [1950], Cartan and Eilenberg [1956], Haefliger [1976], and Fuks [1979].

Corresponding to any nuclear Lie algebra $\mathfrak{g}$ there is a notion of $\mathfrak{g}$ structure on a smooth manifold M. Let A be the ring of C^∞ functions on M, let $v(M)$ be the Lie algebra of smooth vector fields on M, and let $v(M)$ act on $A \otimes_{\mathbb{R}} \mathfrak{g}$ by acting on the first factor.

2.4 Definition. A *smooth differential form of degree q on M with values in* $\mathfrak{g}$ is a continuous A-linear map from the exterior power over A, $\Lambda_A^q v(M)$, to $A \otimes_{\mathbb{R}} \mathfrak{g}$. The *exterior product* and *exterior derivative* are defined as in the finite dimensional case (the latter essentially by formula (2.3)). A $\mathfrak{g}$-*structure* on M is smooth 1-form ω with values in $\mathfrak{g}$ such that

$$d\omega = -\tfrac{1}{2}\omega \wedge \omega.$$

A *concordance* between $\mathfrak{g}$-structures ω_0 and ω_1 is a $\mathfrak{g}$-structure Ω on $M \times [0,1]$ which induces ω_0 on $M \times \{0\}$ and ω_1 on $M \times \{1\}$.

If $f\colon N \to M$ is a smooth map, then $f^*\omega$ is a $\mathfrak{g}$-structure on N, and if $\phi\colon \mathfrak{g} \to \mathfrak{h}$ is a continuous Lie algebra homomorphism, then $\phi \circ \omega$ is an $\mathfrak{h}$-structure on M. These operations are preserved by concordance.

2.5 Example. If $\phi\colon \mathfrak{g} \to v(M)$ is a continuous homomorphism, then it induces a map

$$\phi_A\colon A \otimes_{\mathbb{R}} \mathfrak{g} \to v(M)$$

defined by $\phi(a \otimes g) = a\phi(g)$. If ϕ_A is an isomorphism, then ϕ is called a principal homogeneous $\mathfrak{g}$-space and $\omega = \phi_A^{-1}$ is a $\mathfrak{g}$-structure on M.

2.6 Example. Let G be a topological group whose underlying space is a smooth manifold. Suppose further that the exponential map from the tangent space at the identity to G exists. Then the tangent space at the identity can be identified with the left invariant vector fields on G, and either can be considered as the Lie algebra $\mathfrak{g}$ of G. Furthermore, a $\mathfrak{g}$-structure on M is the same thing as a flat connection in the standard trivial G-bundle $M \times G$. In particular, this example includes the case that G is the group of C^∞ diffeomorphisms of a compact manifold N, so that $\mathfrak{g} = v(N)$, and the case that G is the group of compactly supported diffeomorphisms of any manifold N with Lie algebra $v_c(N)$.

2.7 Example. Consider the Lie algebra $v(N)$, where N is not necessarily compact. Then a $v(N)$-structure on M is equivalent to a smooth foliation on $M \times N$ transverse to the fibers of the projection on M. Indeed, given such a foliation, the value of ω on a tangent vector X to M is found by lifting X to a vector field along the fiber and tangent to the foliation, then taking the component tangent to the fiber. For details, see Haefliger [1978]. A $v_c(N)$ structure corresponds to a foliation which is tangent to $M \times \{x\}$ except for x belonging to some compact set.

2.8 Example. Any C^∞ foliation of codimension m with framed normal bundle gives rise to an $\mathfrak{a}_m^\infty$-structure. Indeed, a framed foliation is defined by a family of 1-forms $\{\omega^i\}$ which satisfy

$$d\omega^i = \omega^j \wedge \omega^i_j$$

for suitably chosen 1-forms ω^i_j. Differentiating this yields 1-forms $\omega^i_{j_1 j_2}$ such that

$$d\omega^i_{j_2} = \omega^k_{j_2} \wedge \omega^i_k - \omega^k \wedge \omega^i_{k j_2}.$$

Continuing in this manner yields the required $\mathfrak{a}_m^\infty$ structure. Indeed, a differential form on M with values in $\mathfrak{a}_m^\infty$ is defined by

$$\omega(X) = \sum_{s=0}^{\infty} \omega^i_{j_1 \dots j_s}(X)\, t^{j_1} \dots t^{j_s} \frac{\partial}{\partial t^i}$$

for any tangent vector X to M. Note that these same calculations, interpreted differently, were used in § II.5 to define cohomology classes associated to a framed foliation. Also, the concordance class of $\mathfrak{a}_m^\infty$-structures here defined depends only on the concordance class of the framed foliation (Fuks [1979, p. 957]). Thus, it is natural to try to define a classifying space for $\mathfrak{a}_m^\infty$-structures and to try to relate it to $B\Gamma_m^\infty$. A number of constructions are in fact possible.

2.9 Definition. A *generalized concordance class* of $\mathfrak{g}$-structures on a topological space X is the assignment to every continuous map f of X into smooth manifold M of a class $\alpha(f)$ of concordant $\mathfrak{g}$-structures on M, such that if $\phi\colon N \to M$ is smooth, then $\alpha(f \circ \phi) = \phi^* \alpha(f)$.

As Fuks [1979, p. 958] has noted, the functor which assigns to every *CW*-complex X the set of generalized concordance classes of $\mathfrak{g}$-structures on X satisfies the hypotheses of Brown [1965]. Hence, there is a *CW* complex $B\mathfrak{g}$ such that generalized concordance classes of $\mathfrak{g}$-structures on X correspond bijectively to homotopy classes of maps into $B\mathfrak{g}$.

2.10 Definition. Let $\mathfrak{g}$ be a nuclear Lie algebra. Then a *k-simplex of* $B\mathfrak{g}$ is a $\mathfrak{g}$-structure on the k-simplex Δ^k.

Face operators can be defined in the usual way, so that the set of k-simplices of $B\mathfrak{g}$ is a semi-simplicial complex whose geometric realization is the space $B\mathfrak{g}$. This construction for $B\mathfrak{g}$ is useful in that $\Delta^k \times M$ is a manifold, so that manifold constructions can be applied to $B\mathfrak{g}$ by applying them to each simplex separately, with appropriate consistency conditions. Thus, there is a de Rham complex for $B\mathfrak{g}$, whose cohomology is isomorphic to the real cohomology of $B\mathfrak{g}$ (Bott, Shulman, and Stasheff [1976]).

2.11 Definition. A differential form ω on $B\mathfrak{g}$ is *smooth under deformation* if, for any smooth family σ_t of k-simplices of $B\mathfrak{g}$ depending on a parameter t in $\mathbb{R}^m$, $\omega(\sigma_t)$ is a smooth family of forms on Δ^k.

Since the exterior derivative of a form smooth under deformation is smooth under deformation, the cohomology of these forms is defined and is called the differentiable cohomology of $B\mathfrak{g}$.

Algebraically, a $\mathfrak{g}$-structure ω has the properties of a flat connection. Hence, the constructions previously given for flat bundles apply immediately to give a homomorphism from $H^*(\mathfrak{g})$ into $H^*(M)$ for any smooth manifold having a $\mathfrak{g}$-structure. In particular, $H^*(\mathfrak{g})$ maps into $H^*(B\mathfrak{g})$, and the image lies in the differentiable cohomology of $B\mathfrak{g}$. Furthermore, if $\mathfrak{g}$ is the Lie algebra of a Lie group G, a $\mathfrak{g}$-structure is a G-trivialized G^δ-bundle, as in the finite dimensional case. Thus $B\mathfrak{g}$ is the homotopy fiber of the map $BG^\delta \to BG$, that is, a kind of homotopy quotient space G/G^δ. Hence, it may also be interpreted (up to homotopy) as the quotient of the singular complex of G by the natural action of G^δ on the right.

Since $\mathfrak{a}_m^\infty$ does not correspond to a group, $B\mathfrak{a}_m^\infty$ cannot be interpreted in the way just described. However, it has been shown that $H^*(\mathfrak{a}_m^\infty)$ is isomorphic to $H^*(v(\mathbb{R}^m))$ (Bott [1975]). The proof uses the homotopy $f_t(x) = tx$ and the Schwartz kernel theorem (§ IV.1). The algebraic analog of this homotopy is the Lie derivative with respect to the radial vector field $x^i(\partial/\partial x^i)$. The decomposition of $C^*(\mathfrak{a}_m^\infty)$ into eigenspaces of this Lie derivative has been used to compute $H^*(\mathfrak{a}_m^\infty)$ algebraically (Gelfand and Fuks [1970]). In order to relate these results to classifying spaces, one considers the construction which associates an $\mathfrak{a}_m^\infty$-structure to a framed C^∞ foliation of codimension m. By looking at the associated foliated microbundle, one extends this construction to $\bar{\Gamma}_m^\infty$-structures on manifolds. Thus there are natural maps

$$H^*(\mathfrak{a}_m^\infty) \to H^*(B\mathfrak{a}_m^\infty) \to H^*(B\bar{\Gamma}_m^\infty).$$

The first is not onto, and it is not known whether the second is an isomorphism. However, $B\bar{\Gamma}_m^\infty$ has the homotopy type of $Bv(\mathbb{R}^m)$ (Fuks [1979, p. 957]). Thus, the composition of the two maps above can be identified with the natural map

$$H^*(v(\mathbb{R}^m)) \to \mathrm{H}^*(Bv(\mathbb{R}^m)).$$

If H is a finite dimensional Lie group whose Lie algebra is contained in $\mathfrak{g}$, then $B(\mathfrak{g}, H)$ can be defined with analogous properties. In particular, the natural map

$$H^*(\mathfrak{a}_m^\infty, O_m) \to H^*(B\Gamma_m^\infty)$$

can be analyzed by using classifying spaces.

The natural map from $v(M)$ to $\mathfrak{a}_m^\infty$ is defined by evaluation at a point for any manifold of dimension m, so that $\mathfrak{a}_m^\infty$ can be viewed as a kind of fiber associated to $v(M)$ as a structure over M. Gelfand and Fuks [1970] have shown that $H^*(\mathfrak{a}_m^\infty)$ is isomorphic to $H^*(Y_m)$, where Y_m is the restriction to the $2m$-skeleton of $BU(m)$ of the universal $U(m)$-bundle. ($U(m)$ is the compact Lie group whose Lie algebra is isomorphic over the complex numbers to that of $GL_m(\mathbb{R})$, and the use of the $2m$-skeleton corresponds to the Bott vanishing theorem.) For any manifold M, there is a bundle $Y_m M$ with fiber Y_m associated to the tangent bundle. It has been shown by Haefliger [1976] that $H^*(v(M))$ is isomorphic to the cohomology of the space of sections of this bundle. The calculation of $H^*(\mathfrak{a}_m^\infty)$ shows that the cohomology is entirely determined by 2-jets, so that there is a natural map

$$H^*(\mathfrak{a}_m^\infty) \to H^*(B\bar{\Gamma}_m^k)$$

for any $k \geqq 2$. This is false for the algebra of formal symplectic vector fields and the symplectic pseudogroup (Gelfand, Kalinin, and Fuks [1972]) and it is unknown whether there is some finite k for which the corresponding result is true.

Much more can be said about the cohomology of Lie algebras, but it would be too great a digression from the main purpose of this monograph. An extensive summary is given by Fuks [1979]. Other references useful in the present context are Bott [1973, 1976], Gelfand, Kazhdan, and Fuks [1972], Godbillon [1973],

Guillemin [1973], Haefliger [1972], Lichnerowicz [1976, 1977], M. Mostow [1976], Schweitzer [1978, Part I], and Trauber [1973].

3. Frobenius Structures

This section deals with the relation between the existence and classification problems for C^∞ foliations and the C^∞ case of the integrability theory discussed in § I.4.

It is clear that if g is a nonzero function, then $g\,df$ defines the same family of submanifolds as df, namely, the components of the level surfaces of f. Thus, from the point of view of foliations of codimension l, it is natural to consider 1-forms ω satisfying one of the two equivalent conditions:

(i) $d\omega \wedge \omega = 0$.

(ii) There exists a 1-form α such that $d\omega = \alpha \wedge \omega$.

Furthermore, there is an equivalent condition in terms of vector fields, namely, the existence of a suitable subalgebra of $v(M)$. However, as soon as singularities are allowed, the various conditions become inequivalent, and the question of a choice of definition becomes important. Since differential forms map contravariantly under any smooth mapping, while vector fields don't map at all except under special conditions, the differential form approach should be taken as fundamental. Furthermore, the two integrability conditions listed above are no longer equivalent, the first being weaker, but more easily verifiable. These ideas, generalized to codimension q, suggest the following definition.

3.1 Definition. A C^∞ *q-Frobenius structure* (or an $\mathscr{F}_q^\infty$-structure) on a manifold M is a C^∞ q-form Ω such that for each $x \in M$, there is a neighborhood U_x and C^∞ differential 1-forms $\omega_x^1, \ldots, \omega_x^q$ on U_x satisfying:

(a) $\Omega \mid U_x = \omega_x^1 \wedge \ldots \wedge \omega_x^q$.

(b) For each α, $d\omega_x^\alpha \wedge \Omega = 0$.

(c) On $U_x \cap U_y$, there exists a C^∞ GL_q-valued function a_β^α such that

$$\omega_x^\alpha = a_\beta^\alpha \, \omega_y^\beta .$$

It is immediate that the a_β^α are the transition functions of a bundle called the normal bundle of the q-Frobenius structure. Moreover, there is a classifying space $B\mathscr{F}_q^\infty$ for $\mathscr{F}_q^\infty$-structures, which can be defined by either of the methods discussed in § 2 for g-structures, namely by the Brown representation theorem or by a semi-simplicial construction.

3.2 Example. Any $\bar{\Gamma}_q^\infty$-structure on a C^∞ manifold gives rise to an $\mathfrak{a}_q^\infty$-structure on the normal microbundle which can be pulled back by the 0-section to an $\mathfrak{a}_q^\infty$-structure on the manifold. The latter structure is given by forms $\{\omega^1, \ldots, \omega^q\}$ which define an $\mathscr{F}_q^\infty$-structure with trivial normal bundle. By analogy with $\bar{\Gamma}_q$, such a structure will be called an $\bar{\mathscr{F}}_q^\infty$-structure. Hence, there are mappings

$$B\bar{\Gamma}_q^\infty \to B\mathfrak{a}_q^\infty \to B\bar{\mathscr{F}}_q^\infty .$$

3.3 Example. Given any Lie algebra $\mathfrak{g}$ of dimension q, a C^∞ $\mathfrak{g}$-structure is an $\bar{\mathscr{F}}_q^\infty$-structure.

3.4 Example. Since any Γ_q^∞-structure gives rise to an $\mathscr{F}_q^\infty$-structure with the same normal bundle, there is a commutative diagram

$$\begin{array}{ccc} B\Gamma_q^\infty & \longrightarrow & B\mathscr{F}_q^\infty \\ & \searrow \quad \swarrow & \\ & BGL_q & \end{array}$$

Thus, the question of when an $\mathscr{F}_q^\infty$-structure is equivalent to a foliation of codimension q can be formulated as follows: Given an $\mathscr{F}_q^\infty$-structure on a C^∞ manifold M and an embedding of the normal bundle into the tangent bundle of M, find a Γ_q^∞-structure on M with the same normal bundle and with the given underlying $\mathscr{F}_q^\infty$-structure. Indeed, if such a Γ_q^∞-structure can be found, it is equivalent to a foliation by the work of Thurston quoted in § 1.

There is another approach to determining when an $\mathscr{F}_q^\infty$-structure is equivalent to a foliation, namely, determining the generic singularities of such a structure and then finding the obstructions to eliminating them. (Every manifold has an $\bar{\mathscr{F}}_q^\infty$-structure, since any q-tuple $\{f^\alpha\}$ of C^∞ functions gives rise to an $\bar{\mathscr{F}}_q^\infty$-structure $\{df^\alpha\}$.) The generic singularities for a Frobenius structure are not the same as those for a mapping, even in codimension 1. Indeed, generically a function has the singularities of a nondegenerate quadratic form, but an integrable 1-form which is not closed may have other types of singularities, generically not isolated, locally obtained from a 1-form in the plane by means of a submersion. (Reeb [1952, p. 141], Kupka [1964]). Moreover, a singular point may have nontrivial holonomy (Moussu [1976, p. 172]). Some related problems associated to the singularities of a foliation have been formulated by Thom [1975].

Whatever method is applied to study global existence problems, some information about the local integrability problem is required. This may be formulated as follows: Given $\{\omega_x^1, \ldots, \omega_x^q\}$ defining an $\mathscr{F}_q^\infty$-structure on U_x, find functions f^α and g_β^α on some neighborhood of x such that

$$\omega^\alpha = g_\beta^\alpha \, df^\beta$$

and the matrix (g_β^α) is nonsingular. This has been solved in various special cases by Moussu [1976] and Malgrange [1976, 1977]. An important step is showing that for certain types of singularities, conditions of the form $d\omega^\alpha \wedge \Omega = 0$ still imply that $d\omega^\alpha$ lies in the ideal generated by $\{\omega^1, \ldots, \omega^q\}$.

Though this discussion has concentrated on C^∞ structures, it is clear that the concept of Frobenius structure makes sense on a C^k manifold, provided $k \geqq 2$. Furthermore, the generic singularities of forms are quite different from those of vector fields, but the latter have been intensively studied by topological dynamicists, and will not be discussed here.

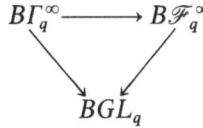

Chapter IV. Metric and Measure Theoretic Properties of Foliations

The first three chapters give background material and relations between foliations and other differential geometric structures. Though current research and open problems are discussed, some of the latter do not seem likely to be solved immediately. In this chapter, the topics discussed are the heart of current activity in the field.

One important idea is that of transverse measure invariant under the holonomy pseudogroup – locally such a measure is the quotient of a measure on the manifold by a measure along the leaves. A transverse invariant measure is related by Poincaré duality to a cycle tangent to the leaves, which exists if the leaves do not grow too fast. The growth of a leaf is measured by the growth of its holonomy pseudogroup, the definition being a natural generalization of the definition of growth of a group.

The introduction of a Riemannian metric is as useful a tool in studying foliations as it is in smooth manifold theory generally. For example, the growth of a leaf can also be measured by the growth of its volume, measured with respect to a Riemannian metric induced on the leaf by a metric on the containing manifold. On a complete manifold, each leaf is also complete. Since a leaf is an immersed submanifold, its second fundamental form is defined, and all the related ideas of submanifold theory can be discussed. Since the plane field orthogonal to the foliation may not be integrable, local submanifold theory must be generalized to any plane field. The difference of the orthogonal projections on the tangent and normal plane fields is a field of linear transformations called the structure tensor of the foliation. When its covariant differential is decomposed according to the invariants of $O(p) \times O(q)$, the second fundamental form and other important tensors appear.

An important special case, the bundle-like metric, is that in which the normal plane field is totally geodesic. Locally, such a foliation has the structure of the level sets of a Riemannian submersion, so is called a Riemannian foliation. The natural parallelization of the principal normal bundle consists of infinitesimal automorphisms of the lifted foliation. Because of this high degree of symmetry, there is a very strong structure theorem if the manifold is compact. If the metric is bundle-like, the associated Γ-structure is Riemannian, so that the usual foliation characteristic classes vanish but a new set can be defined. There is a Laplacian operator that is well-defined on basic forms, and the leaves are minimal submanifolds if and only if the projection onto the normal plane field is an extremal of the appropriate energy functional.

The graph of a foliation is a certain topological groupoid which is foliated by quotients of the fundamental groupoids of the leaves. Associated to the graph there is a C^*-algebra and a version of the Atiyah-Singer index theorem for operators elliptic along the leaves. For a Riemannian foliation the graph is Hausdorff, and if the manifold is compact, then the graph is complete and fibers over the manifold in such a way that the fiber is the holonomy covering of each leaf.

Any C^2 foliation of codimension 1 without holonomy on a compact manifold is topologically equivalent to a smooth foliation with a bundle-like metric. Various theorems about the behavior of germs of C^2 diffeomorphisms of the line are used to generalize this result to a structure theory for such foliations with no condition on the holonomy. The analysis of C^1 germs of diffeomorphisms in any dimension leads to a structure theorem for the neighborhood of a compact leaf with vanishing first Betti number.

If a compact manifold is foliated by compact leaves, the volume of the leaves need not be bounded. However, this situation can be analyzed, and the analysis yields interesting examples.

For a foliation transverse to the fibers of a bundle, the structure is particularly simple if the structure group of the bundle is amenable. The notion of amenability can be generalized to other classes of foliations and interesting results can be obtained. Also, the notion of ergodicity can be defined for foliations. Both of these notions are introduced in the context of measures invariant by a pseudogroup.

For the study of many of these ideas, one needs techniques from functional analysis. Therefore, the first section of this chapter is devoted to summarizing some relevant information in a form suitable for use in the later sections.

1. Analytic Background

The purpose of this section is to summarize some facts about analysis on manifolds which are useful in studying the differential geometry of foliations. One needs to look at vector bundles, such as the tangent bundle or the bundle of p-forms, and to topologize the spaces of C^r sections and of C^r compactly supported sections so that they become complete. An element of the dual space is called a current. For the basic theory of currents, the description will be brief and intuitive, and the reader is referred for details to Dieudonné [E.A., Chaps. 12, 13, 17] and other references. Some topics included are mappings of currents, the weak topology, smoothing operators, and the concrete description of currents in terms of derivatives of continuous sections or of measures. A singular p-chain is a current on the space of smooth p-forms, and a continuous 1-cochain on a nuclear Lie algebra is also a current. The representation of points of a compact convex set in terms of extreme points is discussed briefly. For a foliated manifold, it is necessary to study transverse invariant measures, and more generally, currents invariant under the holonomy pseudogroup. In addition, one must also consider topologies on spaces of C^r foliations.

Since every C^1 manifold is C^1 equivalent to a C^∞ manifold, it is sufficient to establish the theory of currents (also called distributions or generalized functions) for C^∞ manifolds. Then the space of sections of any C^∞ vector bundle admits various differentiation and integration operations. The vector bundle will be assumed to be real unless otherwise stated, but the theory for complex vector bundles is similar. It is convenient to consider initially only C^∞ sections with compact support, since all the operations of interest are defined on such sections. Then the various differential and integral operators to be considered are real-valued linear functions which are continuous with respect to a suitable topology on the space $\mathscr{D}$ of compactly supported C^∞ sections. The topology chosen is the direct (or inductive) limit over compact subsets of the topology of uniform convergence of all derivatives of all orders. (This is called the fine C^∞ topology, the strong C^∞ topology, or the Whitney topology.) Unless the underlying manifold is compact, this topology is not metrizable, but it is locally convex, and $\mathscr{D}$ is complete under this topology in the sense of the theory of uniform spaces. Intuitively, the idea is that under this topology, the limit of compactly supported C^∞ sections is again compactly supported and C^∞. Another interesting topology to consider is the topology of uniform convergence of all derivatives of all orders on compact subsets, also called the coarse C^∞ topology or the weak C^∞ topology. This topology is metrizable, and the completion of $\mathscr{D}$ under the coarse C^∞ topology is the space $\mathscr{E}$ of all C^∞ sections, with no condition on the support of a section. Note that in the case of the tangent bundle of a manifold M, $\mathscr{D}$ is the space $v_c(M)$ and $\mathscr{E}$ is the space $v(M)$ introduced in Definition III.2.1. If $\mathscr{D}^r$ is the space of compactly supported C^r sections and $\mathscr{E}^r$ the space of all C^r sections, then the natural topologies are the fine and coarse C^r topologies respectively. Both these topologies are coarser than the corresponding C^∞ topologies, and $\mathscr{D}$ is dense in $\mathscr{D}^r$ and $\mathscr{E}^r$ in their respective topologies.

The set of continuous linear real-valued functions on any of these spaces is usually indicated by affixing a prime ($'$). In particular, an element of $\mathscr{D}'$ is called a current. Since $\mathscr{D}$ has the finest topology of all the spaces considered, the restriction of a current to $\mathscr{D}^r$, $\mathscr{E}$ or $\mathscr{E}^r$ belongs to $(\mathscr{D}^r)'$, $\mathscr{E}'$, or $(\mathscr{E}^r)'$ respectively. By density, all elements of these latter spaces are thus obtained, so that $(\mathscr{D}^r)'$, $\mathscr{E}'$, and $(\mathscr{E}^r)'$ may be considered as subsets of $\mathscr{D}'$. Thus, the most classical example of a current is a measure, which is an element of $(\mathscr{D}^0)'$.

An important fact about currents is that they are locally defined. This means that if an open covering of the manifold is given, and the value of the current is known for every section whose support is compact and contained in some set belonging to the covering, then the value of the current is completely determined. Thus, the support of a current can be defined, namely, the support is the smallest closed set such that the current vanishes on every compactly supported section whose support is contained in the complement of the given closed set. Furthermore, for a current T with support in a coordinate system with coordinates $(x^1, \ldots, x^m)$, one can define $\partial T/\partial x^i$ by

$$\frac{\partial T}{\partial x^i}[f] = -T\left[\frac{\partial f}{\partial x^i}\right].$$

The minus sign comes from the formula for integration by parts, or Stokes' theorem as it is known to geometers.

1.1 Example. An important class of currents consists of the Dirac measures and their derivatives. Given a point P in M the Dirac measure δ_P is defined by

$$\delta_P[f] = f(P).$$

Its derivatives are given by

$$\frac{\partial^r \delta_P}{\partial x^{i_1} \dots \partial x^{i_r}}[f] = (-1)^r \frac{\partial^r f}{\partial x^{i_1} \dots \partial x^{i_r}}(P).$$

Any current with support at an one point is a finite linear combination of derivatives of Dirac measures and any current of finite order can be approximated by a finite sum of such derivatives.

The elements of $\mathscr{E}'$ are precisely the compactly supported currents, and each such current belongs to $(\mathscr{E}^r)'$ for some finite r. By contrast, it is not true that $\mathscr{D}'$ is the intersection of the spaces $(\mathscr{D}^r)'$. Any compactly supported current is equal to a finite sum of derivatives of measures, and in fact the measures can be assumed to be continuous in the sense that they are obtained as continuous multiples of the volume element of the manifold (Schwartz [1966, p. 91]). For example, on the line, the Dirac measure is the second derivative of the measure $f(x)\,dx$, where

$$f(x) = \begin{cases} 0 & x \leqq 0 \\ x & x \geqq 0. \end{cases}$$

If the condition of continuity of the measure is dropped, then every element of $(\mathscr{E}^r)'$ is the finite sum of derivatives of order at most r of measures. Moreover, in either case, the support of each measure may be assumed to be contained in a preassigned neighborhood of the support of the given current. This description of the structure of compactly supported currents is very useful, for example in the study of $v(\mathbb{R}^m)$ in § III.2.

Frequently, it is convenient to define topologies on spaces of sections by means of a family of seminorms. These differ from the norms that occur in Banach space theory only in that a nonzero element may have a seminorm 0. They are useful because they are the abstract formulation of the estimates that are so much used in analysis. On the other hand, expositions of the theory of currents sometimes avoid any discussion of the topology of $\mathscr{D}$ at all, because it is not a metric topology and there is an adequate equivalent definition of current that avoids mention of the topology of $\mathscr{D}$.

To prove the existence of a current with some desirable property, it is often convenient to look for a fixed point of some mapping defined on a subset of $\mathscr{D}'$, or for a minimum point of some function defined on a subset of $\mathscr{D}'$. For this purpose, a topology with few open sets and therefore many compact subsets is often useful. Thus, the topology of $\mathscr{D}'$ is often taken to be the weak topology, that is, the topology of pointwise convergence. This is the coarsest topology on $\mathscr{D}'$ such that $\mathscr{D}$ can be identified with the set of continuous linear forms on $\mathscr{D}'$, where the identification is made by the natural pairing.

Some very important bundles on a manifold are the cotangent bundle and its exterior powers. The currents on their spaces of sections have been much studied.

1.2 Example. If $\mathscr{D}$ consists of p-forms, then any C^1 singular p-chain c defines a current by

$$c[f]=\int_c f.$$

Furthermore, any $C^1\,(n-p)$-form ϕ defines a current by

$$\phi[f]=\int_M \phi\wedge f.$$

In both these cases, the fact that f is compactly supported is essential. Alternatively, one can require that c be a finite chain or that ϕ have compact support, in which case one gets an element of $\mathscr{E}'$. These ideas can be used in the proof of the Rham's theorem (de Rham [1955]).

1.3 Definition. Given a C^∞ manifold M, $\mathscr{D}_p(M)$ or $\mathscr{D}_p$ denotes the space of compactly supported C^∞ p-forms with the fine C^∞ topology. A *diffuse current* is a smooth $(n-p)$-form, viewed as an element of $\mathscr{D}'_p$. The *boundary* of $T\in\mathscr{D}'_p$ is the current $bT\in\mathscr{D}'_{p-1}$ given by

$$bT[\phi]=T[d\phi].$$

A current is *closed* if $bT=0$ and *exact* if it is in the image of b.

If T is the current defined by a singular chain c, then bT is the current defined by ∂c. In this case, the support of the chain as a distribution is contained in its topological support, but may be properly contained. A C^1 submanifold also defines a current, whose boundary is related to its geometric boundary.

If the manifold is not orientable, then it is useful to introduce also the forms of odd kind. These are sections of the tensor product of one of the bundles just discussed with the orientation bundle of the manifold. On a nonorientable manifold, it is necessary to take ϕ to be an $(n-p)$-form of odd kind in order that it can be interpreted as an element of $\mathscr{D}'_p$. For more details, see the book of de Rham [1955].

For many purposes it is necessary to know how currents behave under smooth bundle mappings. Suppose $p_i\colon E_i\to M_i$ are bundles with the same fiber, and F and f are C^∞ mappings such that the diagram

$$\begin{array}{ccc} E_1 & \xrightarrow{F} & E_2 \\ {\scriptstyle p_1}\downarrow & & \downarrow{\scriptstyle p_2} \\ M_1 & \xrightarrow{f} & M_2 \end{array}$$

in commutative, that is, $f\circ p_1=p_2\circ F$. If f is proper, then there is an induced map

$$\mathscr{D}F\colon \mathscr{D}(M_2)\to\mathscr{D}(M_1)$$

which is continuous with respect to the fine C^∞ topology. Consequently, there is a map

$$\mathscr{D}'F\colon \mathscr{D}'(M_1)\to\mathscr{D}'(M_2)$$

which is continuous with respect to the weak topology. In the case of the bundle of q-covectors, it is important to ask what effect the map $\mathcal{D}'_q F$ has on diffuse currents, where F is the map induced by f. To analyze this question, it is necessary to notice that the definition of diffuse current uses the natural duality structure of the manifold. Indeed, a form of degree q on a manifold of dimension n is made to operate on compactly supported forms of degree $(n-q)$, as an $(n-q)$-chain does. Thus, if $\dim M_i = m_i$, a q-form becomes an $(m_1 - q)$-current, which maps into an $(m_1 - q)$-current, and thus can be diffuse only if it is represented by an $(m_2 - m_1 + q)$-form. If $m_1 < m_2$, the image of a diffuse current is never diffuse, since the image of M_1 under f has empty interior, while the support of a diffuse current has nonempty interior. If $m_1 \geqq m_2$, then it is still necessary to suppose that f is a submersion in order to obtain the result that $\mathcal{D}'_q F$ takes diffuse currents to diffuse currents. This is a considerable restriction, since a proper submersion is a locally trivial fiber space with compact fiber. If a diffuse current is represented by a form ϕ, then $\mathcal{D}'_q F$ lowers its degree by the dimension of the fiber, and $(\mathcal{D}'_q F)(\phi)$ is found by integrating along the fiber. More generally, if the restriction of f to the support of ϕ is proper, $(\mathcal{D}'_q F)(\phi)$ can still be found by integrating along the fiber. If f is not assumed to be proper, there is still induced a continuous map

$$\mathcal{E}F\colon \mathcal{E}M_2 \to \mathcal{E}M_1$$

and the dual map on compactly supported currents. The discussion of diffuse currents proceeds in a manner similar to that for $\mathcal{D}$, since the restriction of f to the support of an element of $\mathcal{E}'$ is certainly proper. The maps $\mathcal{D}'_q F$ and $\mathcal{E}'_q F$, considered as mappings on differential forms of degree $m_1 - q$, induce in cohomology the turn-around homomorphism (Umkehrungshomomorphismus) which is therefore given by integration over the fiber in the interesting cases.

A very useful property of diffuse currents is that they are dense in the weak topology on the space of all currents. This is proved by using smoothing operators which are also useful in a number of other contexts. Classically, a smoothing operator, S on $\mathbb{R}$ is obtained by taking a C^∞ function K on $\mathbb{R}\times\mathbb{R}$ and defining

$$(Sf)(x) = \int_{\mathbb{R}} K(x,y) f(y)\,dy.$$

Sf is a C^∞ function, though f may be very unsmooth, so long as the integral makes sense. In extending this idea to currents on a manifold V, K becomes a C^∞ form on $V\times V$, but there is a technical difficulty which arises in expressing the preceding formula in a coordinate-free way. On the one hand, Fubini's theorem says that $\int K(x,y)\,dx\,dy$ can be evaluated as an iterated integral, and the result is the same whether the first integration is with respect to x or to y. On the other hand, if the variables are interchanged by the change of variable formula, the sign of the integral changes. Thus, one is lead to introduce on any product manifold $V\times W$ the notion of double form. In product coordinates (x^i, y^α), a double form is characterized by being antisymmetric in $\{dx^i\}$ and in $\{dy^\alpha\}$ separately, while dx^i commutes with dy^α. The C^∞ double forms on $V\times W$ which are of degree p on V and degree q on W correspond bijectively to linear continuous maps from $\mathcal{E}'_p(V)$ to $\mathcal{E}_q(W)$. If $W = V$ and the support of the double form K is contained in a neighborhood of the diagonal, then the support of ST defined by

$$ST(x) = \int_V K(x,y)\,T(y)\,dy$$

is contained in a neighborhood of the support of T for any compactly supported current T. In particular, there exists a sequence of double forms $\varrho_k(x,y)$ which define linear continuous maps

$$R_k\colon \mathscr{E}'_p \to \mathscr{D}_{m-p}$$

and linear continuous maps

$$A_k\colon \mathscr{E}'_p \to \mathscr{E}'_{p+1}$$

such that

$$R_k T - T = bA_k T + A_k bT.$$

Furthermore, as the support of ϱ_k shrinks to the diagonal, $R_k T$ converges to T and $A_k T$ converges to 0 in the weak topology on $\mathscr{E}'$. In particular, $R_k T$ is a diffuse current as close to T as one wishes.

Note that the above formula is a cochain homotopy in the sense of algebraic topology. In fact, because of such formulas, smoothing operators play a role in the proof of de Rham's theorem. Another useful class of operators consists of extensions to currents of the usual differential geometric operators on tensor bundles.

In the linear spaces being considered here, the properties of compact convex sets are very important. In particular, it is desirable to represent the elements of such a set in terms of certain special elements, called extreme points. For example, in a finite dimensional simplex, the extreme points are the vertices and every point has a unique representation as a convex linear combination of the vertices. For the unit disk in the plane, the extreme points are the boundary points, and every point has a representation as an integral around the boundary. For a compact convex set K in any locally convex (topological linear) space, an analogous result holds, namely, every point is the barycenter of a probability measure which is supported by the extreme points of K (Choquet [1956] or [1969]). An infinite dimensional example is the set of probability measures on a compact Hausdorff space. This set is compact in the weak (coarse) topology, and the extreme points are the Dirac measures. More interesting is the space of probability measures invariant under a flow on a compact manifold. The extreme points of this space are the ergodic measures. Thus, any invariant measure is built out of ergodic measures, and can be approximated by finite convex combinations of such measures. Instead of restricting to probability measures, one can consider all positive measures. These form a cone, that is, a union of open rays. A cone C in a linear space is said to be compact if there is a continuous linear function L defined on the whole space which is positive on the cone and such that $L^{-1}(1) \cap C$ is compact. Extreme rays of a compact convex cone play the role of extreme points of a compact convex set, and the analogous statements hold.

On a foliated manifold, a very important type of current is the transverse invariant measure. For a foliation of codimension q, this may be viewed as a nonnegative measure on the collection of all q-dimensional transversal submanifolds which is invariant under the holonomy pseudogroup. On the one hand, it is sometimes useful to extend the measure to all Borel transversals, that is, Borel sets which meet each leaf at most countably often. On the other hand, it is sufficient to consider a collection of q-dimensional transversal submanifolds which meets each leaf at least once. There is no natural way to choose such a collection, but a measure on such a collection induces uniquely a measure on any other transversal submanifold. Even though the total measure is not well-defined, the concept of finite total measure is well-defined. Thus, one is lead to consider the cone of all such measures in a suitable linear space. If p is the dimension of the foliation, $\mathscr{D}'_p$ is such a space. Given a transverse invariant measure, it is enough to describe the corresponding current in a foliated coordinate system with transverse disc D and leaves L_y, indexed by $y \in D$. Then for every smooth p-form ω with support in the coordinate system, the value of the current on ω is given by

$$\int_D \left(\int_{L_y} \omega \right) d\mu$$

where μ is the measure on D which represents the transverse invariant measure.

In view of this example, the natural replacement in the theory of currents for the quotient space of a foliation is the space of currents invariant under the holonomy pseudogroup. More generally, one can consider a C^∞ manifold M and a pseudogroup $\mathscr{G}$ of C^∞ diffeomorphisms of M.

1.4 Definition. $\mathscr{D}_p(M/\mathscr{G})$ is the quotient of $\mathscr{D}_p(M)$ by the subspace generated by all elements of the form $\alpha - h^* \alpha$, where $h \in \mathscr{G}$ and the support of α is contained in the range of h. $\mathscr{D}'_p(M)^{\mathscr{G}}$ is the subspace of $\mathscr{D}'_p(M)$ consisting of currents T such that $T[\alpha] = T[h^* \alpha]$.

Since the subspace generated by the elements $\alpha - h^* \alpha$ is not in general closed, $\mathscr{D}_p(M/\mathscr{G})$ is not in general Hausdorff. Dividing by the closure of this subspace yields a Hausdorff space such that $\mathscr{D}'_p(M)^{\mathscr{G}}$ is naturally identified with its dual. Any morphism of pseudogroups induces a continuous linear mapping on each of these spaces, and an isomorphism induces an isomorphism.

In the case of a foliation of codimension q, M is the union of any countable family of q-dimensional transversal submanifolds such that each leaf meets M, and $\mathscr{G}$ is the restriction to M of the holonomy pseudogroup. Up to isomorphism, $\mathscr{D}_p(M/\mathscr{G})$ and $\mathscr{D}'_p(M)^{\mathscr{G}}$ depend only on the foliation, so for a compact total space, M can be taken to be a finite union of transversals of coordinate neighborhoods, and $\mathscr{G}$ is generated by the coordinate changes. An element of $\mathscr{D}'_p(M)^{\mathscr{G}}$ can be called a holonomy invariant p-current. The notion of integration along the fibers can be generalized from fiber spaces to foliations by using invariant forms and currents in the place of forms and currents on the space.

1.5 Proposition. *If $\mathscr{F}$ is a p-dimensional foliation such that the tangent bundle along the leaves is oriented, then there is a continuous open surjective linear map*

$$\int_{\mathscr{F}} : \mathscr{D}_{p+k}(M) \to \mathscr{D}_k(M/\mathscr{G})$$

which commutes with d. Its dual is a linear map

$$\mathscr{D}'_p(M)^{\mathscr{G}} \to \mathscr{D}'_{p+k}(M)$$

commuting with the dual of d. The kernel of $\int_{\mathscr{F}}$ is the vector subspace generated by the $\mathscr{F}$-trivial forms and their exterior derivatives, where a form is $\mathscr{F}$-trivial if it vanishes whenever p of its arguments are tangent to $\mathscr{F}$.

Proof. This proposition is proved by Halfliger [1980]. □

Another place where function space topologies are needed is in the study of deformations of foliations. This is the study of changes in the global structure induced by changes which are small locally, so a topology must be introduced on the set of C^k foliations on a given manifold. Unfortunately, there are two rather natural ways of doing this which are inequivalent if k is finite. One way is to use the C^{k-1} topology on the space of C^{k-1} sections of the appropriate Grassmann bundle. The other is to say that two foliations are C^k close if they can be defined by coordinate changes which are C^k close. The latter method not only generalizes to other pseudogroup structures, but also has the property that if two foliations are C^k close, their holonomy maps are C^k close. In defining these topologies on spaces of foliations, one uses certain neighborhood schemes $(\{\phi_\mu\}, \{K_\mu\})$ and sets of positive real numbers $\{\varepsilon_\mu\}$, almost all zero in the case of the coarse topology. Here ϕ_μ: $V_\mu \to \mathbb{R}^m$ is a coordinate mapping onto an open cube and K_μ is a subset of V_μ which maps onto a closed cube with sides parallel to the coordinate planes. $\{K_\mu\}$ is assumed to be locally finite and $\{\operatorname{int} K_\mu\}$ to be a covering. A foliation lies in the $(\{\phi_\mu\}, \{K_\mu\}, \{\varepsilon_\mu\})$ neighborhood of $\mathscr{F}$ if it admits charts ϕ'_μ: $U'_\mu \to \mathbb{R}^m$ and subsets K'_μ of V'_μ with the above properties, and furthermore satisfying $K_\mu \subset K'_\mu$ and

$$|D^\alpha(\phi'_\mu \phi_\mu^{-1} - \mathrm{id})| < \varepsilon_\mu$$

on $\phi_\mu K_\mu$ for all derivatives D^α of order at most k. The coarse and fine topologies thus defined have the following properties.

1.6 Proposition. *In the coarse and fine C^k topologies, $k \geqq 1$, on the space of C^k foliations of codimension q, the neighborhoods of a given foliation $\mathscr{F}$ defined by using a fixed neighborhood scheme form a basis for the neighborhoods of $\mathscr{F}$. If f is a diffeomorphism of the underlying manifold which is close to the identity in the coarse (respectively fine) C^k topology, then the image foliation $f\mathscr{F}$ is close to $\mathscr{F}$ in the coarse (respectively fine) C^k topology. Furthermore, the holonomy maps vary continuously, in the following sense:*

Let $D(t) = \{x \mid x \in \mathbb{R}^q$ and $|x| < t\}$. Let h: $[0,1] \times D(1) \to M$ be a C^k map such that $h \mid \{t\} \times D(1)$ is an embedding transverse to $\mathscr{F}$ for each $t \in I$ and such that for each $x \in D(1)$, $h([0,1] \times \{x\})$ lies on a single leaf of $\mathscr{F}$. Then for any $\mathscr{G}$ sufficiently close to $\mathscr{F}$ in the coarse C^k topology, there is a C^k map k: $[0,1] \times D(\frac{1}{2}) \to M$ such that

(i) $k \mid \{0\} \times D(\frac{1}{2}) = h \mid \{0\} \times D(\frac{1}{2})$
(ii) *for each* $x \in D(\frac{1}{2})$, $k([0,1] \times \{x\})$ *lies on a single leaf of* g
(iii) $k(\{t\} \times D(\frac{1}{2})) \subseteq h(\{t\} \times D(\frac{3}{4})$ *and* $k \mid \{t\} \times D(\frac{1}{2})$ *is an embedding for each* t
(iv) *for each* $t \in [0,1]$, g *is transverse to* $h(\{t\} \times D(\frac{3}{4}))$
(v) k *is near to* $h \mid [0,1] \times D(\frac{1}{2})$ *in the coarse C^k topology.*

If $\mathscr{G}$ is close to $\mathscr{F}$ in the fine C^k topology, then $\{k_\mu\}$ exist simultaneously for any family $\{h_\mu\}$ such that $\{\operatorname{Im} h_\mu\}$ is locally finite family.

Proof. The proof is given by Epstein [1977]. □

This completes the overall discussion of the analytic background for the study of foliations. Certain additional topics will be discussed as they are needed.

2. Measure, Volume, and Foliations

Fubini's theorem applied to a foliated coordinate neighborhood gives a natural decomposition of the volume into tangent and quotient volumes. In this section, the global structures associated to this local decomposition will be studied. In the case of the quotient measure, the global analog is the notion of transverse invariant measure which was introduced in § 1. Geometric consequences can be drawn from the existence of a nonzero transverse invariant measure. The global analog of the tangent volume is the total volume of a leaf. For a foliation of a compact manifold with compact leaves, the important question is whether the total volume of the leaves is bounded. In general, the volume of a leaf is infinite, but the rate of growth of the volume is an important invariant. This rate of growth can also be expressed as a property of the holonomy pseudogroup. If a leaf has nonexponential growth, then its closure supports a transverse invariant measure, and a converse holds in codimension 1. The holonomy groupoid is also an interesting geometric object, sometimes known to geometers as the graph of the foliation. It is locally compact, and has associated to it a C^*-algebra which generalizes the convolution algebra of integrable functions on a locally compact group. These and a number of related items will be discussed in this section.

For many purposes, the plane field tangent to an oriented p-dimensional foliation is conveniently described as a section of a bundle with fiber the space of oriented p-planes in euclidean space $\mathbb{R}^m$, where m is the dimension of the manifold. Since this fiber is not a vector space, it is more convenient now to view an oriented p-plane as a ray in the exterior power $\Lambda_p \mathbb{R}^m$. A volume structure on the manifold yields an isomorphism for each x from $\Lambda_p T_x M$ to $\Lambda_{m-p} T_x^* M$, the map being given by the interior product, and the inverse map by the interior product with the m-vector structure dual to the volume structure. Since in approximating a foliation by a smooth structure, the rays at nearby points are averaged, it is natural to consider the more general situation in which a cone (a union of rays) is given in each $\Lambda_p T_x M$. A cone C in a locally convex (topological linear) space is said to be compact if there is an element L of the dual space such that $L(x) > 0$ for $x \in C$ and $L^{-1}(1) \cap C$ is compact. If the locally convex space is metric, each such L can also be used to define neighborhoods of C in the set of all compact cones. Namely, C' is close to C provided that $L^{-1}(1) \cap C'$ is compact and close to C in the Hausdorff metric on compact subsets of the given locally convex space. The same definition applies to cones in different fibers of a trivialized bundle.

2.1 Definition. A *cone structure* on a closed subset X of a C^r manifold M, $r \geqq 1$, is a family of compact, convex cones $\{C_x\}$ such that C_x lies in $\Lambda_p T_x$ and varies continuously with x. A differential p-form ω is *positive on the cone structure* if $\omega(v) > 0$ for each $v \in C_x$ and all $x \in X$.

Note that positive p-forms exist, since for each x, ω_x is a linear functional on $\Lambda_p T_x$ and can be identified with the functional L in the definition of "compact cone".

As has already been recalled in § 1, finite sums of derivatives of Dirac currents are dense in the space $\mathscr{D}'_r$ of all r-currents. Since the analog of Dirac currents is easy to define for cone structures, it is convenient to turn this idea into the definition of the associated structure currents. Throughout the discussion, the weak topology on $\mathscr{D}'_p$ will be assumed.

2.2 Definition. Given a cone structure $\{C_x\}$ on X, a *Dirac current* for the structure is a current of the form $T(\omega) = \omega(v_x)$, where $v_x \in C_x$ for some $x \in X$, and ω is a p-form. A *structure current* is an element of the smallest closed convex set in $\mathscr{D}'_p$ containing the derivatives of the Dirac currents for the cone structure. A *structure cycle* is a closed structure current, and a *structure boundary* is an exact structure current. A structure current is called a *foliation current* if the cone at each point consists of a single ray.

If $D_I = \partial^r/\partial x^{i_1} \ldots \partial x^{i_r}$ is a partial derivative operator with respect to some coordinate system, then for any Dirac current T

$$(D_I T)(\omega) = (-1)^r (D_I \omega)(v_x)$$

is the expression for $D_I T$ with respect to this coordinate system.

2.3 Proposition. *The set of all structure currents is a closed convex cone in $\mathscr{D}'_p$. If X is compact, the structure currents of order 0 form a compact convex cone $\mathscr{C}^0$, and each such current has a representation*

$$\int_X v \, d\mu$$

where μ is a nonnegative measure on X and v is a μ-integrable function such that $v(x) \in C_x$. Furthermore, the structure cycles and structures boundaries form closed convex subcones of the corresponding cone of currents.

Proof. The structure currents of order 0 are those that can be approximated by finite sums of undifferentiated Dirac currents. These are the structure currents in the sense of Sullivan [1976a], and his proof applies. The last statement follows from the facts that the boundary operator is continuous and linear, and that the boundaries are defined by the vanishing of a finite number of periods. □

Though any current can be approximated by diffuse currents, in the case of a structure current, the approximating diffuse currents are not usually structure currents. To obtain the best results possible in this direction, it is necessary to choose carefully the kernel K_ε of the diffusion operator. K_ε is a closed form Poincaré dual to the diagonal in $M \times M$. It can be so chosen that as ε approaches 0, K_ε approaches the integral current of the diagonal in the topology of m-currents on $M \times M$, the support of K_ε is contained in the ε neighborhood of the diagonal, and the dual m-vector approaches the oriented m-vector determined by the tangent planes to the diagonal. Let D_ε denote the diffusion operator corresponding to K_ε.

2.4 Proposition. *D_ε is a linear operation on currents which commutes with ∂ and preserves homology classes. For structure currents of order 0, $D_\varepsilon T$ is a diffuse current relative to a slightly larger cone structure, and the support of T is contained in the interior of the support of $D_\varepsilon T$.*

Proof. See the paper of Sullivan [1976a]. □

In order to see what properties foliation currents should have, one may look at the case of a nonsingular vector field on a compact manifold. It has been known for a long time that in this case, the closure of every leaf contains the support of a nonzero measure on the manifold invariant by the flow defined by the vector field. Dividing locally by the measure along the orbits inherited from the natural measure on the group $\mathbb{R}$ gives a transversal invariant measure with the same support. This division can also be viewed as the interior product by the vector field, so is related to the foliation currents defined by the vector field. For higher dimensional foliations, a compact leaf still gives rise to a transversal invariant measure, but other leaves may not. In view of these facts, one proceeds by determining the local structure of a foliation current in order then to fix the relation between foliation currents and transversal invariant measures. Foliations with all leaves compact are then studied before the problem of existence of transversal invariant measures is taken up in general.

Consider therefore a flat coordinate neighborhood for the foliation. The coordinates will be denoted as usual by (x, y), and it will be assumed that they are consistent with a fixed unit p-vector structure, so that

$$v = \partial/\partial x^1 \wedge \ldots \wedge \partial/\partial x^p$$

is well-defined, and the value of any foliation current is completely determined by its value on this particular v. A derivative of arbitrary order with respect to x will be called tangent and denoted by D_x, while one with respect to y will be called transverse and denoted by D_y, though this splitting depends upon the choice of coordinates. Indeed, upon changing from (x, y) to another set of foliation coordinates (ξ, η), the derivative $\partial^{r+s}/\partial x^I y^J$, where $|I| = r$ and $|J| = s$, goes into

$$\sum_{0 \leqq t \leqq s} a^{KL}\, \partial^{r+s}/\partial\xi^{k_1} \ldots \partial\xi^{k_{r+t}}\, \partial\eta^{l_1} \ldots \partial\eta^{l_{s-t}},$$

so that the terms with the maximal number of transverse derivatives determine each other, but no other terms can be considered in isolation. Thus, it is convenient to obtain a representation of the structure of a foliation current with respect to the product structure of a particular coordinate neighborhood, and then later to study the effect of the change to some other product structure consistent with the foliation structure. Consider therefore a fixed coordinate neighborhood, and let $B^p \times B^q$ be a product of compact convex sets contained in the coordinate neighborhood, such that $L_y = B^p \times \{y\}$ is contained in a leaf. Let W be a convex open neighborhood of $B^p \times B^q$ such that $\bar{W}$ is compact and contained in the coordinate neighborhood. Any derivative D of order r of any Dirac foliation current is given by the formula

$$(DT)(\omega) = (-1)^r (D\omega_{1\dots p}),$$

so that by regarding $\omega_{1\dots p}$ as an element of $\mathscr{D}_0$, the structure theory for $\mathscr{D}'_0$ may be applied. Since the restriction of a foliation current T to W has compact support, hence finite order k, it can be written as the finite sum of derivatives of order at most k of measures with support in $\bar{W}$. Thus, for any p-form ω with support in W, $T(\omega)$ is the finite sum of terms like

$$\int_{\bar{W}} (D\omega_{1\dots p})\, d\mu.$$

If ω has support in $B^p \times B^q$, then so does D_ω, and therefore the integration can be taken over $B^p \times B^q$ with respect to the induced measure. This induced measure can be disintegrated along the local projection map for the foliation, so that the preceding expression becomes

$$\int_{B^q} \left(\int_{L_y} D\omega_{1\dots p}\, d\mu'_y \right) d\mu''$$

where μ'' is a measure on B^q and μ'_y is a measure with support in L_y. Let $D = D_x D_y$ be the decomposition of D into tangent and transverse parts. Then standard results about the differentiation of currents which depend upon a parameter [Schwartz 1966, p. 105] imply that the preceding formula can be rewritten

$$\int_{B^q} D_y \left(\int_{L_y} D_x \omega_{1\dots p}\, d\mu'_y \right) d\mu''. \tag{2.5}$$

2.6 Proposition. *In a local product neighborhood, the value of any foliation current T on a p-form ω with support in $B^p \times B^q$ is a finite sum of terms, each given by formula* (2.5), *where the notation is defined in the preceding paragraph. If the current is a cycle and has order* 0, *then on the interior of $B^p \times B^q$*

$$T = \int_{B^q} (L_y)\, d\mu''$$

where (L_y) is the current of integration on L_y and μ'' is nonnegative and unique. For this result, the foliation must be at least C^1.

Proof. The preceding paragraph contains the proof of the first statement, as well as the necessary definitions. The second statement is proved by Sullivan [1976a]. The idea is to study the current $[L_y]$ defined by

$$[L_y](\omega) = \int_{L_y} \omega_{1\dots p}\, d\mu'_y.$$

Because T is a cycle, the support of $\partial [L_y]$ does not meet the interior of $B^p \times B^q$ for almost all y in the interior of B^q. Thus, on the interior of such an L_y, $[L_y]$ is a constant multiple $f(y)$ of the integration current (L_y), where $f(y)$ is Borel measurable and the measure $f(y)\,\mu''$ depends only on T. It is nonnegative because a foliation current of order 0 is nonnegative and disintegrates into nonnegative measures. The differentiability hypothesis makes sure that C^1 functions $h(y)$, which are dense in the continuous compactly supported functions on B^q, are pulled back by the local projections to functions $k(y)$ such that $k(y)\omega$ is in the domain of $\partial[L_y]$ for every C^∞ $(p-1)$-form ω. Note that the last formula is a sharpening of Proposition 2.3. ▯

The preceding proposition gives a local description of each foliation current. In particular, a foliation cycle of order 0 is described in terms of a transversal current, the measure μ''. Globally, such a foliation cycle is a current invariant under the holonomy pseudogroup, as described in the discussion following Definition 1.4. It is convenient to reformulate the description in terms of compact neighborhoods, as follows.

2.7 Definition. Let (S, W) denote a pair consisting of a foliation coordinate neighborhood W containing $B^p \times B^q$ as above, and a compact convex set $S = \{x_0\} \times B^q$. A family $\{(S_\mu, W_\mu)\}$ of such pairs with the further property that each leaf meets the interior of some S_μ will be called a *covering family*. A holonomy invariant current is then represented locally by a current T_μ defined on the interior of S_μ, which admits an extension to some neighborhood of S_μ in

$$\{(x, y) \mid (x, y) \in W, x = x_0\}.$$

A holonomy invariant measure is said to have *finite total mass* if the measure of $\bigcup_\mu \operatorname{int}(S_\mu)$ is finite for some suitably chosen covering family of pairs.

Because of the uniqueness in Proposition 2.6, the measure on the quotient of a flow box contained in another flow box is related to the measure on the larger quotient by the natural coordinate change between the two flow boxes. Thus, with a slight change of notation, the foliation cycle gives rise to a transversal invariant measure, which will have finite total mass if the manifold is compact.

2.8 Theorem. *For an oriented C^1 foliation of a compact manifold, there is a canonical one-one correspondence between foliation cycles of order* 0 *and transversal invariant nonnegative measures of finite total mass. Any compact leaf is the support of a foliation cycle.*

Proof. The construction of a transversal invariant measure from a foliation cycle has already been described. Conversely, given such a measure, choose a finite covering family $\{(S_\mu, W_\mu)\}$ such that $\{\operatorname{int}(B^p \times B^q)_\mu\}$ is a covering of the manifold, and let $\{\phi_\mu\}$ be a partition of unity subordinate to the covering. Define

$$T(\omega) = \sum_\mu \int_{S_\mu} \int_{L_y} \phi_\mu \omega \, d\mu''.$$

To see that this is independent of the choice of covering and partition of unity, compare a given covering with a refinement, using the uniqueness of μ''. For details, see Ruelle and Sullivan [1975]. T is closed since

$$T(d\omega) = \sum_\mu \int_{S_y} \int_{L_y} d(\phi_\mu \omega) \, d\mu''$$

and the inner integral vanishes because $\phi_\mu \omega$ vanishes on the boundary of L_y. Clearly, each construction is the inverse of the other, so the first statement is proved. The second then follows from the fact that a compact leaf meets each S_μ in a finite number of points. □

One version of the Poincaré duality theorem asserts that the intersection number between cycles in complementary dimensions induces a nonsingular pairing in homology with rational coefficients. Consequently, the cohomology class dual to the class of a given cycle is evaluated on some other cycle by taking the intersection number of the two cycles. The transverse invariant measure makes a similar statement possible for foliation cycles, using real coefficients.

2.9 Definition. A q-dimensional manifold with boundary is *transverse* to a foliation of codimension q if the injection map can be extended to some manifold containing N in its interior so as to be transvere to each leaf.

Suppose N is a compact q-dimensional manifold, possibly with boundary, transversal to an oriented foliation of codimension q, such that the boundary does not meet the support of some foliation cycle T. Then the corresponding transversal invariant measure induces a measure μ on N, and the total mass $\mu(N)$ is the intersection number of T and N, in the sense of de Rham's theory of currents. Alternately, one can diffuse T to a closed q-form ϕ with support near the support of T, and obtain the same number by integrating ϕ over N. Since ϕ vanishes near the boundary, it defines a class in the q-th cohomology group of N relative to the boundary. On the other hand, since $H_q(M;\mathbb{R})$ has a basis consisting of the homology classes defined by certain embedded submanifolds (Thom [1954]), the cohomology class dual to T is given by the periods of ϕ on these submanifolds. It would be interesting to put these submanifolds into general position with respect to the foliation, and thereby to relate them geometrically to the support of T. Unfortunately, this seems to require knowledge of the generic structure of a smooth map from $\mathbb{R}^q$ to $\mathbb{R}^q$, which knowledge seems likely to remain unavailable for some time, at least for large q.

By Proposition 2.3, the foliation currents of order 0 form a compact convex cone $\mathscr{C}^0$ in the space $\mathscr{D}'_p$. Furthermore, the foliation cycles and boundaries form compact convex subcones $\mathscr{Z}^0$ and $\mathscr{B}^0$ respectively. By identifying the transversal invariant nonnegative measures of finite total mass with these cycles, one can obtain various interesting results about the set of all these measures. Since $\mathscr{D}_p$ is the dual space of $\mathscr{D}'_p$, a linear functional on $\mathscr{D}'_p$ whose kernel meets $\mathscr{C}^0$ only at the vertex can be interpreted as a differential p-form which is positive on the foliation, hence defines a volume element on each leaf. By using these ideas and the Hahn-Banach theorem, Sullivan [1976a] obtains the following results.

2.10 Proposition. *Let $\mathscr{F}$ be an orientable p-dimensional C^1 foliation of a compact manifold. If there is no nonzero transversal invariant measure of finite total mass for $\mathscr{F}$, then there is an exact p-form positive on the leaves of $\mathscr{F}$. If there is no closed p-form positive on the leaves of $\mathscr{F}$, then there is a nonzero transversal invariant measure of finite total mass such that the corresponding foliation cycle bounds. If there are both nonzero transversal invariant measures of finite total mass and closed p-forms positive on the leaves, then the natural map from invariant measures to* *homology classes is proper and the image is a compact cone* $\mathbb{C}$ *in* $H_p(M; \mathbb{R})$. *Furthermore, the interior of the dual cone* $\mathbb{C}'$ *in* $H^p(M; \mathbb{R})$ *consists of the cohomology classes of the closed forms positive on* $\mathscr{F}$.

Another way of studying all the foliation cycles is by looking at the union of their supports. This method lead to generalization of certain results in the study of dynamical systems.

2.11 Definition. The *Poincaré recurrence set* $P(\mathscr{F})$ of an oriented foliation $\mathscr{F}$ is the union of the supports of all foliation cycles. The foliation is said to be *totally recurrent* if the Poincaré recurrence set is the whole space.

In the theory of dynamical systems, there are various theorems showing that points in this set always come back close to themselves after arbitrarily large times. (For example, see Nemytskii and Stepanov [1960, pp. 447–459].) There are also interesting results for foliations of codimenion 1 which will be given below. Moreover, the Poincaré recurrence set has some quite strong stability properties.

2.12 Proposition. *Suppose that M is compact and* $\mathscr{F}$ *is* C^1. *Then*

(i) $P(\mathscr{F})$ *is a closed invariant set.*

(ii) *If* $P(\mathscr{F})$ *is empty and the tangent planes of* $\mathscr{F}'$ *are* C^0 *close to those of* $\mathscr{F}$, *then* $P(\mathscr{F}')$ *is empty.*

(iii) *If* $P(\mathscr{F})$ *is nonempty,* $\mathscr{N}$ *is any neighborhood of* $P(\mathscr{F})$, *and the tangent planes of* $\mathscr{F}'$ *are sufficiently close topologically to those of* $\mathscr{F}$, *then the support of any foliation cycle of* $P(\mathscr{F}')$ *meets* $\mathscr{N}$.

Proof. If $\{C_i\}$ is a sequence of foliation cycles which is dense in the cycles of mass 1 (with respect to a given family of transversals), then $\sum_{i=1}^{\infty} 2^{-i} C_i$ is a foliation cycle whose support is $P(\mathscr{F})$. The last two statements follow from the existence of an exact form positive on any compact set contained in the complement of $P(\mathscr{F})$. □

The simplest example of a totally recurrent foliation is a compact foliation.

2.13 Definition. A *compact foliation* is a foliation of a compact manifold such that every leaf is compact.

By Theorem 2.8, a compact foliation is totally recurrent. A simple example of such a foliation is a fiber space with compact fiber and base. A more interesting example has for leaves the orbits of the S^1 action ϕ on S^3 given by

$$\phi(t, z_1, z_2) = (\exp(2\pi ipt) z_1, \exp(2\pi iqt) z_2)$$

where (z_1, z_2) is a pair of complex numbers such that $z_1 \bar{z}_1 + z_2 \bar{z}_2 = 1$, t is a real number modulo 1, and (p, q) is a pair of relatively prime integers. This is very close to being a fiber space, except that the holonomy groups of the leaves $z_1 = 0$ and $z_2 = 0$ are cyclic of orders q and p respectively. Also, considered as curves in S^3, the leaves have lengths which vary continuously except on the two exceptional leaves, where the lengths of nearby leaves are approximately q and p times as great. Similar remarks hold for a locally free action of any compact group. In order to study more general compact foliations, it is convenient to restrict to C^1 foliations of manifolds with empty boundary, though many of the fundamental facts are true more generally (see Epstein [1976]). Then any Riemannian metric on the manifold induces on each leaf a Riemannian metric. Since each leaf is compact, it has finite volume, but it is not necessarily the case that the volumes are bounded, as they are in the simple examples. Understanding the structure of compact foliations requires analyzing the behavior of these volumes.

2.14 Definition. The *volume function* for a compact C^1 foliation $\mathscr{F}$ on a manifold M with empty boundary is the function V on M such that $V(x)$ is the volume of the leaf through x.

2.15 Proposition. *Let X be a locally compact saturated subset of M, x a point of X, n a positive integer, and ε a positive real number. Then there is a neighborhood U of x in X such that for every y in U, either $V(y) > nV(x)$ or there is an integer j, $1 \leqq j \leqq n$, such that*

$$|V(y) - jV(x)| < \varepsilon.$$

In particular, the restriction V_X of V to X is lower semicontinuous. The set of points where V_X is continuous and the set of points where it is locally bounded are open and dense in X. V_X is continuous at x if and only if the holonomy group at x of the induced foliation is trivial. V_X is bounded on some neighborhood of x if and only if this holonomy group is finite. When this holds, there is a neighborhood W in X of the leaf through x such that

(i) *each leaf in W has arbitrarily small saturated neighborhoods.*
(ii) *The quotient of W by the foliation is Hausdorff, and the quotient map is closed.*
(iii) *The holonomy group in X of each leaf is isomorphic to a subgroup of $O(q)$ acting orthogonally on a subset of D^q, where q is the codimension of the foliation.*

Proof. See Edwards, Millett, and Sullivan [1977, § 4] and Epstein [1976]. □

The simple example above of an S^1 action on S^3 illustrates a number of these possibilities, and also provides some insight into the proof. If Proposition 2.15 is applied with $X = M$, it produces a dense open subset W_c on which V is continuous and a dense open subset W_b on which V is bounded. The proposition may then be applied to W_c to show that the foliation of W_c is a locally trivial fibration, and also to W_b to show that the foliation is a more general object in which the local product structure is replaced by the local structure of a flat bundle over a leaf. The proposition may also be applied to the complement of each of these open sets, to begin an inductive process for analyzing unbounded behavior.

2.16 Definition. The *Epstein hierarchy* for a compact foliation of M is the collection of subsets X_α of M indexed by the ordinal numbers $\alpha \geqq 1$ such that

(i) $X_1 = \{x \in M\colon V \text{ is not bounded at } x\}$.
(ii) $X_{\alpha+1} = \{x \in X_\alpha\colon V|X_\alpha \text{ is not continuous at } x\}$.
(iii) If β is a limit ordinal, $X_\beta = \bigcap_{\alpha<\beta} X_\alpha$.

There are a number of obvious variations, but this definition is the one used by Edwards, Millett, and Sullivan.

2.17 Proposition. *If $\beta > \alpha$, then X_β is closed and nowhere dense in X_α. If X_1 is not empty, then there is a countable ordinal α such that X_α is not empty, but $X_{\alpha+1}$ is empty. For each leaf L, there is an ordinal β such that L is contained in X_β and does not meet $X_{\beta+1}$.*

Proof. See Edwards, Millett, and Sullivan [1977, p. 19], where all but the last statement are proved. The last statement follows from the fact that the smallest ordinal μ such that X_μ does not meet L cannot be a limit ordinal because each X_α is saturated. □

From the general theory given in § 1, it is clear that it is important to know what the extreme rays of the cone $\mathscr{Z}^0$ of order 0 foliation cycles are. In the case of a compact foliation, they are easy to describe.

2.18 Proposition. *The nonnegative multiples of the integration cycle of a compact leaf form an extreme ray in the cone $\mathscr{Z}^0$. For a compact foliation, there are no other extreme rays.*

Proof. A compact leaf gives rise to a Dirac measure, hence to an extreme ray. Conversely, given an extreme cycle, consider the smallest μ so that X_μ meets the support of the given cycle. Each leaf in $X_\mu - X_{\mu+1}$ has finite holonomy and arbitrarily small invariant neighborhoods in X_μ. Hence, the transverse measure is a Dirac measure of a single leaf. □

A closer examination of the behavior of leaves near X_1 makes it possible to prove that in some cases X_1 is empty. At the same time, it gives a clue on the construction of examples with X_1 nonempty.

2.19 Proposition. *If X_1 is nonempty, then there exists a homotopy*

$$h_t\colon L \to M - X_1, \quad 0 \leqq t < \infty$$

such that

(i) *each h_t is an embedding onto a leaf with trivial holonomy.*

(ii) *Given any neighborhood of X_1, there is a t_0 such that for $t \geqq t_0$, $h_t(L)$ is contained in the given neighborhood.*

(iii) *For integer values n of t*

$$\lim_{n\to\infty} V(h_n(L)) = \infty.$$

Proof. See Edwards, Millett, and Sullivan [1977, p. 22]. The idea is that the set of leaves with trivial holonomy is a dense open subset of M whose complement in $M - X_1$ is a countable union of smooth submanifolds of codimension at least 2. Hence, each component N_i of $M - X_1$ contains many families h_t satisfying (i). To satisfy (iii), it is necessary to choose N_i so that V is unbounded on N_i. Since the number of components may be infinite, the existence of such an N_i is not obvious, and must be proved by appealing to the arguments used to show that pointwise periodic transformations are periodic. To find h_t with all the required properties, one must approach X_1 along a single end of N_i. $\square$

Proposition 2.19 is sometimes referred to as the Moving Leaf Proposition. In some cases, it enables one to prove the existence of a bound for the volume function.

2.20 Proposition. *Suppose that a compact foliation is such that the homology classes of the leaves lie in an open half-space in the vector space $H_p(M; \mathbb{R})$, where p is the dimension of the leaves. Then there is a bound on the volume of the leaves, so that X_1 is empty and the structure of the foliation is as described in Proposition 2.15.*

Proof. By Choquet's theorem (Choquet [1956] or [1969]) any foliation cycle T is a weighted average of extreme points of the cone $\mathscr{Z}^0$. In other words, there is a measure v on the set E of integration cycles of leaves such that for any p-form ω

$$T(\omega) = \int_E \left(\int_L \omega \right) dv.$$

By hypothesis, there is a closed form ω whose integral over each leaf is positive, so that for each cycle T, $T(\omega) > 0$, and therefore no foliation cycle bounds. By Proposition 2.10, there is a closed form ω positive on the leaves, that is, positive on the tangent plane to each leaf at each point. Consider the integral of ω over $h_t(L)$. On the one hand, this integral is constant, since the leaves are homotopic and therefore homologous. On the other hand, ω is a multiple of the volume element of the leaves by a function which is bounded away from 0. Hence, the integral over $h_t(L)$ is unbounded as t goes to ∞, a contradiction. $\square$

It remains to discuss the examples where X_1 is nonempty. It is easy to prove that it is always empty in codimension 1. In codimension 2, it is also always empty, but the proof is not easy (Edwards, Millett, and Sullivan [1977] and Vogt [1976]). One must construct a smoothly embedded open 2-manifold which is transverse to the foliation, closed in some open neighborhood of X_1, and meets each leaf in X_1. This leads to a contradiction by a homological argument with the same geometric content as the proof of Proposition 2.20. In codimension 3 and above, X_1 can be nonempty. Epstein and Vogt [1978] have constructed an analytic compact foliation of dimension 1 on a 4-manifold with X_1 a 3-complex. Sullivan [1976b] constructed the first example with X_1 nonempty, a curve family on a 5-manifold, with X_1 a 4-manifold. Sullivan's paper also describes a similar example, due to Thurston, which is analytic. Vogt [1977] has found examples such that the Epstein hierarchy is infinite. In order to give some idea of the phenomenon of unbounded volume, Thurston's example will now be given.

Let N be the 3-dimensional nilpotent Lie group consisting of the matrices

$$\begin{pmatrix} 1 & x & z \\ 0 & 1 & y \\ 0 & 0 & 1 \end{pmatrix}, \qquad (x, y, z) \in \mathbb{R}^3.$$

A basis for the left invariant forms is

$$\{dx, dy, \eta = dz - x\,dy\}.$$

Let (α, β) be coordinates on the abelian Lie group $\mathbb{R}^2$, and let Γ be the subgroup of $N \times \mathbb{R}^2$ consisting of all points with integer coordinates. The quotient of $N \times \mathbb{R}^2$ by the left action of Γ is a compact analytic manifold M and the analytic vector field

$$\sin(4\pi\beta)\cos(2\pi\alpha)\frac{\partial}{\partial x} + \sin(4\pi\beta)\sin(2\pi\alpha)\frac{\partial}{\partial y} + 2\sin^2(2\pi\beta)\frac{\partial}{\partial \alpha} + \left(x\sin(4\pi\beta)\sin(2\pi\alpha) - \frac{1}{2\pi}\cos^2(2\pi\beta)\right)\frac{\partial}{\partial z}$$

on $N \times \mathbb{R}^2$ is invariant by Γ, hence defines an analytic vector field X on M. Note that $(N \cap \Gamma)\backslash N$ is a nontrivial circle bundle over T^2, so that M is the product of this circle bundle with another copy of T^2.

2.21 Proposition. *The orbits of X are all compact and β is constant on each. If β is a multiple of $\frac{1}{2}$, the orbits are the fibers of the natural fibering over T^3 of the 4-manifold defined by holding β constant. In terms of the natural parameter defined by X on its integral curves, each such fiber has length 2π. For other values of β, the length is $(2\sin^2(2\pi\beta))^{-1}$, hence is unbounded for β near an integer multiple of $\frac{1}{2}$.*

Proof. The differential equations corresponding to this vector field can be solved explicitly, and the necessary facts read off from the solutions. Sullivan [1976b] gives a geometric discussion of this and a number of other examples, which in particular explains where these formulas come from. ▯

This completes the discussion of compact foliations. If the leaves are not compact, the situation is more complicated, since there may be no foliation cycles, even if the manifold is compact. For example, for a foliation defined by the action of a Lie group on a compact manifold, there is an invariant measure supported on the closure of any leaf, but the volume of the leaf may grow so rapidly that the transverse measure produced by dividing by it vanishes. Thus, for any foliation, it is necessary to study the growth rates of a leaf. This depends only on the foliation, and small growth rate implies the existence of a foliation cycle supported on the closure. In the C^1 case, the growth will be studied by means of an auxiliary Riemannian metric, then the results will be extended to the C^0 case by introducing the idea of growth of the holonomy pseudogroup.

For a C^1 foliation, a Riemannian metric on the manifold induces one on each leaf, considered as an immersed submanifold. Let $D(x,r)$ be the set of points at a distance less than r from x with respect to the induced metric in the leaf through x. Since $D(x,r)$ is an open set with compact closure, it has a finite volume with respect to the induced volume element.

2.22 Definition. The *growth function* G_x of the foliation at x is the function which has for its value at r the volume of $D(x,r)$ for $r > 0$. The leaf L_x through x has *exponential growth* if

$$\liminf_{r\to\infty} \frac{1}{r} \log G_x(r) > 0.$$

Otherwise, it has *nonexponential growth.* L_x has *polynomial growth of degree d* if there is a polynomial of degree d in r which is an upper bound for $G_x(r)$.

In order to work with this notion, a second definition is helpful.

2.23 Definition. A *bounded volume form* on a Riemannian manifold is a multiple of the unit volume form by a function which is both bounded and bounded away from 0. A complete Riemannian manifold is *not closed at* ∞ if there exists a bounded volume form ω and a bounded form η such that $d\eta = \omega$.

2.24 Proposition. *If a leaf of a C^1 foliation is not closed at ∞, then it has exponential growth of volume. If it is closed at ∞, then there is a transverse invariant measure whose support is contained in its closure.*

Proof. The proof of this proposition is deferred until § 3, where foliations of Riemannian manifolds are considered in more detail. These results are included in Propositions 3.3 and 3.6. □

The concept of growth of a pseudogroup is a direct generalization of the notion of growth of a group. It is defined relative to a given generating set.

2.25 Definition. A pseudogroup Γ of homeomorphisms of a topological space X is said to be *generated by the subset* S if each element of Γ can be obtained from the elements of S by a finite number of pseudogroup operations (composition, inversion, restriction, union). Γ is *finitely* (*countably*) generated if S is *finite* (*countable*). S is *symmetric* if the inverse of each element in S belongs to S.

2.26 Definition. Let Γ be a pseudogroup of homeomorphisms of the topological space X, S a finite symmetric generating set for Γ, and x a point of X. Let $\Gamma^n(x)$ be the set of points which can be reached from x by the composition of at most n elements of S. Then the *growth function* of Γ at x (with respect S) is the function g_x such that $g_x(n)$ is the cardinality of $\Gamma^n(x)$. Γ has *exponential growth* at x if

$$\liminf_{n\to\infty} \frac{1}{n} \log g_x(n) > 0.$$

Otherwise, it has *nonexponential growth* at x. Γ has *polynomial growth of degree d* at x if $g_x(n)$ is dominated by a polynomial of degree d in n. If Γ is countably generated, it has *nonexponential growth* at x provided it can be written as the increasing union of finitely generated pseudogroups, each of which has nonexponential growth at x.

2.27 Proposition. *Let Γ be a finitely generated pseudogroup. Then the orbit $\Gamma(x)$ is finite if and only if Γ has polynomial growth of degree* 0 *at x. If Γ has polynomial* *growth at x, it has nonexponential growth at x. If X is compact metric and Γ has nonexponential growth at x, then there exists a regular Borel measure μ on X such that*

(i) *For every $\gamma \in \Gamma$ and every Borel set $A \subset \mathscr{D}(\gamma)$, $\mu(\gamma(A)) = \mu(A)$.*
(ii) $\mu(X) = 1$.
(iii) *The support of μ is contained in the closure of $\Gamma(x)$.*

If Γ is countably generated and the rest of the hypotheses are satisfied, then μ with the preceding properties exists. If the action of Γ is equicontinuous at x and X is compact, then μ with the preceding properties exists.

Proof. An orbit is finite if and only if $g_x(n)$ is constant for large n, which is equivalent to being bounded since g_x is monotone and integer valued. The second statement follows immediately from the definitions. The last two sentences are proved by Plante [1975, pp. 334–336]. The idea is to construct an invariant functional on the space of continuous functions by averaging each function over the sets $\Gamma^n(x)$. Actually, the averaging is done over a subsequence $n(j)$ such that

$$\lim_{j \to \infty} \frac{\operatorname{card}(\Gamma^{n(j)+1}(x) - \Gamma^{n(j)-1}(x))}{\operatorname{card} \Gamma^{n(j)}(x)} = 0.$$

Such a subsequence exists because the growth is nonexponential. A countably generated pseudogroup is treated as the union of finite generated pseudogroups. The notion of equicontinuity for a pseudogroup is defined and the existence of a measure proved by Sacksteder [1965, p. 85]. □

Consider now a C^0 foliation of a compact manifold. In order to study the growth of the holonomy pseudogroup, it is necessary to construct a specific generating system. Let $\{U_\mu\}$ be a finite covering by flat coordinate neighborhoods such that $\bar{U}_\mu$ is compact and contained in a flat neighborhood V_μ, and each plaque of V_μ meets at most one plaque of V_ν if $\nu \neq \mu$. The existence of such a covering follows from an argument of Reeb [1952, p. 103]. Let X be the union of the local quotients X_μ of the coordinate neighborhoods. Then the transition functions form a symmetric generating set for a pseudogroup Γ acting on X, and Γ determines the full holonomy pseudogroup for the foliation.

2.28 Definition. The *growth function* $g_x(n)$ of the leaf through $x \in U_\mu$ is the growth function of the pseudogroup Γ at the point of X_μ corresponding to x. In particular, the leaf has *exponential, nonexponential,* or *polynomial growth* if the pseudogroup does.

2.29 Proposition. *For a C^1 foliation, the two definitions of growth of a leaf are equivalent, and independent of the choices involved. A leaf is compact if and only if it has polynomial growth of degree* 0. *A leaf with polynomial growth has nonexponential growth.*

Proof. The equivalence is proved by Plante [1975, pp. 338–339]. The rest is clear from the preceding discussion. □

It is now possible to prove the existence of transversal invariant measure in the C^0 case, provided the growth is slow enough. Furthermore, in codimension 1, a partial converse holds.

2.30 Theorem. *Let L be a leaf with nonexponential growth of a C^0 foliation of a compact manifold. Then there exists a nontrivial finite transversal invariant measure for the foliation which has support contained in the closure of L.*

Proof. See Plante [1975, p. 339]. One uses the pairs of coordinate neighborhoods $\{U_\mu \subset V_\mu\}$ described above. A measure exists on the union of the quotients of the $\bar{U}_\mu$ by Proposition 2.27. This measure must be extended to the union of the quotients of the V_μ. In V_μ, any Borel set with compact closure can be written as the finite union of sets, each of which is mapped by a holonomy element into some set in the interior of one of the $\bar{U}_\mu$. Thus, its measure can be defined to be the sum of the measures of the image sets. This measure is then extended to all Borel sets. ▯

2.31 Corollary. *Let $\mathscr{F}$ be a transversally oriented codimension q foliation of class C^k, $k \geqq 1$, of a compact manifold M. Let N be a compact q-manifold without boundary transverse to $\mathscr{F}$, and suppose there is a leaf L with nonexponential growth which meets N. Then N represents a nonzero element of $H_q(M;\mathbb{R})$.*

Proof. The foliation cycle carried by the closure of L has nonzero intersection number with N. ▯

As the first example, consider the case of foliations of dimension 1 of a compact manifold. The only 1-manifolds are the circle and the line, which have polynomial growth of orders 0 and 1 respectively. Hence, the closure of each leaf supports a transversal invariant measure. If the foliation is oriented and X is a tangent vector field, then any measure μ on the manifold defines a foliation current T according to the formula

$$T(\omega) = \int \omega(X)\, d\mu.$$

Furthermore, the extreme rays of the cone of invariant measures are the ergodic measures, that is, the measures such that the only invariant subsets are those of measure 0 and 1. Hence, any invariant measure can be approximated by a finite convex combination of ergodic measures. Also, for each invariant measure μ, the corresponding foliation cycle is approximated by averaging over larger and larger pieces of the orbit through x, for μ-almost all x (Schwartzman [1957]).

For foliations of dimension greater than 1, the leaves can have very large growth rates. Nonetheless, in codimension 1, there is enough restriction on the relationships among leaves to make some interesting extensions of the theory possible. The main result is a restriction on the growth rate for leaves in the support of a transversal invariant measure.

2.32 Theorem. *Let M be a compact manifold, $\mathscr{F}$ be a transversally oriented codimension one foliation of M of class C^k, $k \geqq 0$, and μ be finite transversal invariant* *measure for $\mathcal{F}$. Then there is a homomorphism Φ from $\pi_1(M)$ into the reals which depends only on μ, $\mathcal{F}$, and M, such that any leaf L contained in the support of μ has polynomial growth of degree at most*

$$\max\{0, p(\mu) - 1\}$$

where $p(\mu)$ is the rank of the image of Φ, considered as a free abelian group.

This theorem has many interesting consequences. Since $p(\mu)$ is at most equal to the rank $b_1(M)$ of the first homology group, if $b_1(M)$ is at most 1, the existence of a compact leaf is implied. In general, if there exists a leaf of nonexponential growth, then there exists a leaf of polynomial growth of a rather low degree. On the other hand, Hector [1977] has given examples of a C^∞ 2-dimensional foliation of a compact 3-manifold having leaves of polynomial growth of every degree, leaves of growth larger than polynomial but less than exponential, and leaves of exponential growth.

Proof of Thereom 2.32. This theorem is proved by Plante [1975, pp. 345–349]. For $k \geqq 1$ and any codimension q, Φ is essentially the element of $H^q(M; R)$ associated to μ by Poincaré duality and Theorem 2.8. If Φ is viewed as a homomorphism from $H_q(M; \mathbb{R})$ to $\mathbb{R}$, the image of Φ has a basis $\{\xi_i | i = 1, \ldots, p(\mu)\}$. Φ can then be considered as a homomorphism from $\mathbb{R}^{p(\mu)}$ to $\mathbb{R}$ which takes the unit point ε_i on the i-th coordinate axis to ξ_i. The number of lattice points $\sum n_j \varepsilon_j$ such that

$$|\Phi(\sum n_j \varepsilon_j)| < \delta$$

is bounded by a polynomial of degree $p(\mu) - 1$ in $n = \sum |n_j|$ because the volume of the intersection of $\Phi^{-1}((-\delta, \delta))$ with a ball of radius r is bounded by a polynomial of this degree in r. The hypothesis of codimension 1 is used to show that the plaques of L lying in a given coordinate neighborhood are distinguished from each other by the measure of the transverse segment from a base point, hence that the number of plaques that can be reached by a chain of fixed length cannot grow faster than the image of Φ. Also, for $q = 1$, Φ can be defined in a way which applies to the case $k = 0$ as well. Indeed, any loop can be deformed to one consisting of segments along the leaves and segments transverse to the leaves, and then the value of Φ is the sum of the signed measures of the transverse segments, the sign being chose according to the orientation of the segment. Plante also gives an example to show that in codimension greater than 1, the support of an invariant measure can contain leaves of exponential growth, so the restriction on codimension is essential. □

Clearly, the $p(\mu)$ in Theorem 2.32 cannot be large unless the first Betti number of the manifold is large. However, there are further restrictions on the fundamental group beyond those immediately implied by this remark.

2.33 Proposition. *If $p(\mu) \geqq 2$, then $\pi_1(M)$ has a free subgroup of rank at least 2, and its commutater subgroup has a free group as a quotient.*

This result is proved by Plante [1975], as is the following generalization of the Reeb stability theorem (Reeb [1952]).

2.34 Proposition. *Let L be a compact leaf of a C^0 foliation of codimension 1 of a compact manifold. Suppose $H^1(L; \mathbb{R}) = 0$ and $\pi_1(L)$ has nonexponential growth. Then the manifold is a bundle over S^1 having the leaf as fiber.*

This proposition is false for C^0 foliations without the restriction on $\pi_1(L)$, but for C^1 foliations this restriction is not needed (Theorem 5.1).

A useful property of foliations of codimension 1 is that any leaf which is not a closed set is cut by a transversal circle. Thus, if the leaf lies in the support of a foliation cycle, it has positive intersection number with the closed transversal. These remarks prove that a foliation cycle which is not supported on compact leaves does not bound. This and a number of other interesting applications of the general theory to codimension 1 foliations are given by Sullivan [1976a]. Some of these applications also depend on another simple useful property, namely, that a form of degree 1 or $n-1$ is always decomposable, and therefore defines a plane field of the appropriate dimension.

The preceding discussion has shown the close relation beteen transversal invariant measures and the growth rate of individual leaves. In the special case of a foliation transverse to the fibers of a bundle with compact fiber, transversal invariant measures can be shown to exist if the group of the bundle is not too large. One possible meaning of the phrase "not too large" is given by the following definition.

2.35 Definition. A discrete group G is said to be *amenable* if there is a nontrivial left invariant finitely additive probability measure defined on the set of all subsets of G.

For a discussion of this concept see for example the book of Greenleaf [1969]. The class of amenable groups is closed under formation of subgroups, quotient groups, direct limits, and group extensions. Any finite or solvable group is amenable, as is any group of nonexponential growth. For present purposes, the property of interest is that an amenable group acting on a compact space leaves invariant a Borel measure of total measure 1.

2.36 Proposition. *Let $p: E \to M$ be an orientable C^1 bundle with compact fiber and discrete amenable structure group. Then the natural foliation transverse to the fibers admits a transversal invariant measure, and the cohomology map*

$$p^*: H^*(M; \mathbb{R}) \to H^*(E; \mathbb{R})$$

is injective with a multiplicative left inverse. In particular, this applies if the fundamental group of the base is amenable. If the fiber is a circle, then there exist leaves of polynomial growth.

Proof. This and various related theorems are proved by Hirsch and Thurston [1975]. The idea is that the invariant Borel measure on the fiber gives rise to a transversal invariant measure for the foliation. Integration along the fiber with respect to this measure provides the required left inverse to p^*. An alternative proof of injectivity can be based on the fact that the fiber has positive intersection number with any foliation cycle, hence cannot bound in E. □

The concepts of amenability and ergodicity, useful in the study of measures invariant under a group action, can be generalized to transverse measures invariant under the holonomy pseudogroup. Thus, on a compact manifold, if the holonomy pseudogroup acts amenably with respect to a sufficiently good transverse invariant measure and almost every leaf has nonpositive curvature then almost every leaf is flat (Zimmer [1983]). If it acts ergodically and the leaves are noncompact Riemannian symmetric spaces, then in most cases the Riemannian structure of the leaves is determined up to isometry by the equivalence class of the transverse invariant measure (Zimmer [1980]). Both of these results are related to theorems about manifolds in which a hypothesis about the fundamental group replaces a hypothesis about the transverse measure. Thus, a compact manifold with amenable fundamental group and nonpositive curvature is flat, and a compact quotient of a symmetric space of noncompact type is determined up to isometry by its fundamental group. Results can also be obtained about foliations with leaves of curvature that is negative but bounded way from 0 (Zimmer [1982]).

The holonomy groupoid of a foliation, defined in § I.5, is very useful in studying the measure theory of the foliation. This groupoid is primarily a measure of the behavior near a leaf, but the leaves in general are not independent, and there are global properties of holonomy. Already mentioned is the fact that the set of leaves with trivial holonomy is residual (Proposition I.5.7.). Furthermore, the set of germs of local submersions is a covering with respect to the leaf topology (Lemma I.5.5). Neither of these facts gives much of a picture of the relations among nearby leaves. However, understanding of these relations is necessary for both the geometric and the analytic theory of foliations. It is immediate from the definitions that the holonomy groupoid of a leaf is isomorphic to the quotient of the fundamental groupoid by the relation that two homotopy classes α and β with the same endpoints are equivalent if $\alpha\beta^{-1}$ induces the identity germ. This same remark applies to a foliated manifold, provided the manifold is taken with the leaf topology. Thus, before studying the holonomy groupoid of a foliation, it is convenient to recall the properties of the fundamental groupoid of a topological space, and to study its quotient groupoids in the case of a connected space.

As a set, the fundamental groupoid $\Gamma(X)$ of the topological space X is the quotient of the set of paths in X by the relation of homotopy with fixed endpoints. It is a groupoid under the composition and inversion induced by the composition and inversion of paths. The topology of $\Gamma(X)$ is the quotient topology of the compact open topology on the space of paths, and with this topology, $\Gamma(X)$ is a topological groupoid. Let A: $\Gamma(X) \to X$ take each path class to its initial point, while Z takes each path class to its final point. Let S: $X \to \Gamma(X)$ take the point x to the constant path at x. Let π denote the group $\pi_1(X, x_0)$ relative to some base point x_0, and let $\tilde{X}$ denote the universal covering space of X. Furthermore, let $Z \oplus A$ denote the restriction to the diagonal of $Z \times A$.

2.37 Proposition. *Let X be connected, locally arcwise connected, and semilocally simply connected. Then $Z \oplus A$: $\Gamma(X) \to X \times X$ is a covering isomorphic to the covering corresponding to the diagonal subgroup $\Delta(\pi)$ of $\pi \times \pi$, considered as the fundamental group of $X \times X$ and operating on the left of $\tilde{X} \times \tilde{X}$. A and Z are* *fiberings of $\Gamma(X)$ over X with fiber $\tilde{X}$ and section S, but in general these fiberings cannot be trivialized in such a way that the image $S(X)$ remains fixed.*

Proof. If $\tilde{X}$ is regarded as the space of path classes beginning at x_0, then π operates on the left of $\tilde{X}$. There is also an isomorphic action on the right obtained by reserving the direction of all paths. Thus, the product of two copies of the left action is isomorphic to the action of $\pi\times\pi$ given by

$$(\alpha,\beta)\cdot(\tilde{x},\tilde{y}^{-1})=(\alpha\cdot\tilde{x},\tilde{y}^{-1}\cdot\beta^{-1}).$$

By the usual arguments used in the construction of the universal covering space, $Z\times A$ is a covering projection. If $Z(\tilde{x})=x$ and $Z(\tilde{y})=y$, then

$$Z\oplus A(\tilde{x},\tilde{y}^{-1})=(x,y).$$

Any given path class from y to x can be written in the form $\tilde{y}^{-1}\tilde{x}$ for suitable choices of $\tilde{x}$ and $\tilde{y}$, and then the set of all path classes from y to x can be written

$$\{\tilde{y}^{-1}\gamma\tilde{x}\,|\,\gamma\in\pi_1(X,x_0)\}.$$

Furthermore $\tilde{y}^{-1}\beta^{-1}\alpha\tilde{x}=\tilde{y}^{-1}\tilde{x}$ if and only if $\alpha=\beta$, so that in terms of the left $\pi\times\pi$ action on $\tilde{X}\times\tilde{X}$, $\pi_1(\Gamma(x),s(x_0))$ appears as the subgroup

$$\Delta(\pi)=\{(\alpha,\alpha)\,|\,\alpha\in\pi\}.$$

By standard arguments, A (respectively Z) is a fibering, and the fibers are clearly the sets of path classes with fixed initial point (respectively endpoint). Since $S(X)$ is a lifting of the diagonal in $X\times X$, if X is a manifold then the tube around $S(X)$ is equivalent to the tangent bundle of X, so the indicated trivializations are certainly impossible without the assumption of parallelizability. This completes the proof. □

Because X is arcwise connected, corresponding to any normal subgroup N in $\pi_1(X,x_0)$, there is a well-defined normal subgroup in the fundamental group at any other point of X. Thus, it makes sense to introduce an equivalence relation on $\Gamma(X)$ by considering α and β to be equivalent provided $\alpha\beta^{-1}$ belongs to the normal subgroup at the appropriate point. The quotient of $\Gamma(X)$ by this equivalence relation will be denoted by $\Gamma_N(X)$. As a first step in understanding $\Gamma_N(X)$, it is convenient to consider the corresponding algebraic construction.

2.38 Definition. Let π be a group and N a normal subgroup. Then the *graph of* N is the subgroup of $\pi\times\pi$ defined by

$$\Gamma_N(\pi)=\{(\alpha,\beta)\,|\,\alpha\beta^{-1}\in N\}.$$

In particular $\Gamma_\pi(\pi)=\pi\times\pi$ and $\Gamma_1(\pi)=\Delta(\pi)$.

2.39 Proposition. *There is a split exact sequence*

$$1\to N\xrightarrow{h}\Gamma_N(\pi)\underset{\Delta}{\overset{k}{\rightleftarrows}}\pi\to 1$$

where $h(\beta)=(1,\beta)$, $k(\alpha,\beta)=\alpha$, *and* $\Delta(\alpha)=(\alpha,\alpha)$. *There is also an exact sequence*

$$1\to N\times N\xrightarrow{i}\Gamma_N(\pi)\xrightarrow{j}\pi/N\to 1$$

where $i(\alpha,\beta)=(\alpha,\beta)$ *and* $j(\alpha,\beta)=N\alpha$. *The normalizer* $(\Gamma_N(\pi))$ *contains* $\Gamma_{CN}(\pi)$, *where* C *is the center, but may be larger than this group or smaller than* $\pi\times\pi$.

Proof. The equation

$$(\alpha,\beta)=(\alpha,\alpha)(1,\alpha^{-1}\beta)$$

holds in the group $\pi\times\pi$. Clearly, the first factor belongs to the image of Δ and the second factor to the image of h. This establishes a bijection between pairs (α,β) and pairs $(\alpha,\gamma)\in\pi\times N$ where $\gamma=\alpha^{-1}\beta$. In terms of the latter representation, the group operation is given by

$$(\alpha_1,\gamma_1)(\alpha_2,\gamma_2)=(\alpha_1\alpha_2,\alpha_2^{-1}\gamma_1\alpha_2\gamma_2).$$

From this, all the required properties of the first sequence can be verified, while the properties of the second sequence are easily obtained directly. Also, direct calculation suffices to verify that $\Gamma_N(\pi)$ is normal in $\Gamma_{CN}(\pi)$. Also, by direct calculation, $(\Gamma_1(\pi))$ is $\Gamma_C(\pi)$, which will be smaller that $\pi\times\pi$ unless π is abelian. For an example in which the normalizer is larger than $\Gamma_{CN}(\pi)$, let π be the alternating group A_4 and let N be the normal subgroup of order 4. Since conjugation by an even permutation preserves not only N but each of its cosets, $\Gamma_N(A_4)$ is normal in $A_4\times A_4$. On the other hand, the center of A_4 consists of the identity only, so $\Gamma_{CN}(A_4)=\Gamma_N(A_4)$ and is therefore not the entire normalizer. The completes the proof. □

2.40 Definition. Let X be connected, locally arcwise connected, and semilocally simply connected, and let N be a normal subgroup of $\pi=\pi_1(X,x_o)$. Then the *graph* $\Gamma_N(X)$ *of the covering of* X *with fundamental group* N is the quotient of the fundamental groupoid $\Gamma(X)$ be the relation that α is equivalent to β if and only if $\alpha\beta^{-1}$ belongs to the subgroup corresponding to N in $\pi_1(X,x_1)$, where x_1 is the initial point of α.

2.41 Proposition. *$\Gamma_N(X)$ is homeomorphic to the quotient of $\tilde{X}\times\tilde{X}$ by the left action of $\Gamma_N(\pi)$, considered as a subgroup of $\pi\times\pi$.*

$$Z\oplus A\colon \Gamma_N(X)\to X\times X$$

is the covering corresponding to the inclusion of $\Gamma_N(\pi)$ into $\pi\times\pi$, and there is a covering $\Gamma(X)\to\Gamma_N(X)$ corresponding to the inclusion of $\Delta(\pi)$ in $\Gamma_N(\pi)$. In general, neither of these coverings is regular. $\tilde{X}\times(\tilde{X}/N)$ is a regular covering of $\Gamma_N(X)$ with automorphism group π, which acts by the diagonal action. $\tilde{X}/N\times\tilde{X}/N$ is a regular covering of $\Gamma_N(X)$ with automorphism group π/N. Z is a fibering of $\Gamma_N(X)$ over X with fiber $\tilde{X}/N$ and a section induced by the section of Z considered as a map on $\Gamma(X)$. The action of π in the fundamental group of the fiber is given by $\alpha\cdot n=\alpha n\alpha^{-1}$.

Proof. The isomorphism obtained in Proposition 2.37 gives rise to the homeomorphism between $\Gamma_N(X)$ and the quotient of $\tilde{X}\times\tilde{X}$ by $\Gamma_N(\pi)$. This isomorphism also gives rise to the covering mentioned in the second sentence. By Proposition 2.39, the subgroups in question need not be normal and therefore the coverings in general are not regular. The first exact sequence of Proposition 2.39 describes the covering of $\Gamma_N(X)$ by $\tilde{X}\times\tilde{X}/N$, while the second exact sequence describes the covering of $\Gamma_N(X)$ by $\tilde{X}/N\times\tilde{X}/N$. Combining the first exact sequence with the fact that Z fibers $\Gamma(X)$ over X leads to the rest of the statements in this Proposition. □

Finally, the preceding constructions can be applied to the study of foliations. Let M be a manifold of dimension m with a foliation $\mathscr{F}$ of dimension p.

2.42 Definition. The *graph* $\Gamma(\mathscr{F})$ *is a quotient of the fundamental groupoid of* M, taken with the leaf topology. Two path classes α and β with the same endpoints are equivalent if $\alpha\beta^{-1}$ has trivial holonomy. $\Gamma(\mathscr{F})$ may also be called the *holonomy groupoid of* $\mathscr{F}$.

M with the leaf topology is not connected, but the above theory applies to each component L (leaf) separately, giving a manifold of dimension $2p$ of the form $\Gamma_N(L)$, where N is the kernel of the holonomy homomorphism of L. However, in addition $\Gamma(\mathscr{F})$ has a locally euclidean topology of dimension $m+p$ by means of which the graphs of the leaves are related.

2.43 Definition. A foliation is of class $C^{k,l}$ if the derivatives of order at most $(k-l)$ along the leaves of the coordinate transformations are of class C^l.

2.44 Proposition. *Let M be a connected, Hausdorff, m-dimensional manifold of class C^k, $k \geqq 0$, admitting a countable basis for its topology. Let $\mathscr{F}$ be a foliation of M of dimension p and class C^l, $l \leqq k$. Then the graph $\Gamma(\mathscr{F})$ is a connected, $(m+p)$-dimensional, σ-compact manifold of class C^l having closed points and admitting a countable basis. If the graph is Hausdorff, then it is paracompact, normal, and metrizable. If every holonomy germ associated to every closed path is conjugate to an analytic germ, then $\Gamma(\mathscr{F})$ is Hausdorff. $\Gamma(\mathscr{F})$ admits a $C^{k,l}$ foliation of dimension $2p$ whose leaves are the graphs of the leaves of $\mathscr{F}$. The leaf topology of $\Gamma(\mathscr{F})$ is paracompact and Hausdorff. $\Gamma(\mathscr{F})$ also admits a pair of transverse, p-dimensional, $C^{k,l}$ foliations tangent to the $2p$-dimensional foliation. If A and Z map a path class to its initial and final points respectively, then they are continuous, open C^l submersions of $\Gamma(\mathscr{F})$ onto M, and $Z\oplus A\colon \Gamma(\mathscr{F})\to M\times M$ is an immersion whose image is the (graph of the) relation making two points equivalent if and only if they lie on the same leaf. The restriction of Z to any $2p$-dimensional leaf $\Gamma_N(L)$ induces an isomorphism between the holonomy groups of $\Gamma_N(L)$ and of L. The map S which takes each point into the homotopy class of the corresponding constant path is a C^l embedding of M into $\Gamma(\mathscr{F})$, and the map T which interchanges the initial and final points of each path class is a C^l involution which preserves the $2p$-dimensional foliation and interchanges the two p-dimensional foliations. The groupoid operations are differentiable of class C^l.*

Proof. For a connected Hausdorff manifold, the notions of paracompactness, σ-compactness, and admitting a countable basis are equivalent. Moreover, if a manifold has all these properties, then so does each leaf of any foliation of it. A construction of coordinates for $\Gamma(\mathscr{F})$ has been given by Connes [1979, pp. 112–114]. The idea is to choose flat coordinate systems $\{(t,u)\}$ and $\{(t',u)\}$ at the beginning and end of a path having the same transverse coordinates u. Then $\{(t',t,u)\}$ is a coordinate system on the graph, and the changes of coordinates have the form

$$\Phi(t',t,u) = (\phi'(t',u), \phi(t,u), \psi(u)).$$

This atlas exhibits the required foliations and differentiability properties. In terms of coordinates,

$$\begin{aligned} Z(t',t,u) &= (t',u) \\ A(t',t,u) &= (t,u) \\ S(t,u) &= (t,t,u) \\ T(t',t,u) &= (t,t',u) \end{aligned}$$

which shows that these maps have the required differentiability and rank. In terms of coordinates, the composition in the holonomy groupoid is given by

$$(t'',t',u)(t',t,u) = (t'',t,u)$$

and the inverse is given by T, so the groupoid operations are of class C^l. If N is the holonomy kernel of L, then there is a natural map λ_L of $\Gamma_N(L)$ into $\Gamma(\mathscr{F})$ which is continuous and open with respect to the $2p$-dimensional leaf topology on $\Gamma(\mathscr{F})$, so the image is a connected open subset. This set is maximal connected, since Z is continuous with respect to the leaf topologies and maps the image of λ_L into the maximal connected set L. Hence, the leaves of the $2p$-dimensional foliation of $\Gamma(\mathscr{F})$ are each C^k diffeomorphic to one of the sets $\Gamma_N(L)$ as required. $\Gamma(\mathscr{F})$ is connected with respect to the $(m+p)$-dimensional topology, since $S(M)$ is connected and meets each leaf of the $2p$-dimensional foliation. Two points P and Q can be separated if either $Z(P) \neq Z(Q)$ or $A(P) \neq A(Q)$. However, each fiber of Z over a point of L is a covering space of L whose automorphisms are given by the holonomy group of L. It follows that each point is closed. Moreover, each equivalence class of the equivalence relation generated by the nonseparation relation is contained in a fiber of the covering, so is a discrete set in either p-dimensional leaf topology. The graph is Hausdorff if and only if each germ belonging to the holonomy group at P which is the identity on some open set whose closure contains P is the identity germ at P. In particular, this condition holds if each holonomy germ is conjugate to an analytic germ. Assuming that M is paracompact, Epstein, Millett, and Tischler [1977] have constructed a countable collection of homeomorphisms such that all holonomy maps are obtained as germs of these mappings. Thus, the topology of the graph is obtained from a countable family of coordinate neighborhoods, each of which admits a countable basis for its topology. Hence, $\Gamma(\mathscr{F})$ admits a countable basis, and is therefore σ-compact. If $\Gamma(\mathscr{F})$ is Hausdorff, it is paracompact. Each leaf of the original foliation has a countable basis, so by the preceding argument applied to its foliation of codimension 0, $\Gamma(L)$ has a countable basis. Since $\pi_1(L)$ is countable,

$\Gamma_N(L)$ has a countable basis for all N. Thus, in all the leaf topologies, $\Gamma(\mathscr{F})$ is the free union of paracompact Hausdorff manifolds, and therefore is paracompact. The coordinate formulas show that A and Z are submersions and open maps, while the characterization of the image of $Z \oplus A$ follows immediately from the definitions. If $\pi = \pi_1(L)$ and N is the holonomy kernel of L, then the first exact sequence of Proposition 2.40 is part of the homotopy sequence of the fibering obtained by restricting Z to a leaf $\Gamma_N(L)$. In particular, $Z_\# = k$, so $Z_\#$ induces an isomorphism between $\Gamma_N(\pi)/N \times N$ and π/N. Since the holonomy of $\Gamma_N(L)$ is trivial along $N \times N$, the required holonomy isomorphism is produced. S is clearly one-one, and the coordinate expression shows that it is an immersion. It remains to prove that S is open. However, $\Delta: M \to M \times M$ is open when considered as a map onto its image, $A \oplus Z$ is continuous, and $S = (A \oplus Z)^{-1} \circ \Delta$. T is a diffeomorphism because it is differentiable and $T = T^{-1}$. This completes the proof of Proposition 2.44. □

The following examples illustrate some important classes of graphs.

2.45 Examples. Consider the compact foliation defined by the action

$$\phi(t, z_1, z_2) = (\exp(2\pi i p t) z_1, \exp(2\pi i q t) z_2)$$

of S^1 on S^3 already discussed in connection with Definition 2.13. The graph of this foliation can be described by the orbits of an $S^1 \times S^1$ action on $S^3 \times S^1$ as follows:

$$\phi(t_1, t_2, z_1, z_2, z_3) = (\exp(2\pi i p t_1) z_1, \exp(2\pi i q t_1) z_2, \exp(2\pi i t_2) z_3)$$

$$z_1 \bar{z}_1 + z_2 \bar{z}_2 = 1, \quad z_3 \bar{z}_3 = 1,$$

p and q relatively prime.

In general, for a locally free, effective group action, the graph is obtained by forming the Cartesian product with the left action of the group on itself.

2.46 Example. Let ϕ be a C^∞ function from $\mathbb{R}$ to itself such that:

$$\begin{aligned} &\phi(y+1) = \phi(y) \\ &\phi(y) = 0 \quad \tfrac{1}{2} \leqq y \leqq 1 \\ &\phi(y) > 0 \quad 0 < y < \tfrac{1}{2}. \end{aligned}$$

Then the one-form ω on $\mathbb{R}^2$ defined by

$$\omega = dy - \phi(y) \sin 2\pi y \, dx$$

defines a foliation of $T^2 = \mathbb{R}^2/\mathbb{Z}^2$. The leaves $y = c$ for $\frac{1}{2} \leqq c \leqq 1$ cover compact leaves. The leaves $y = \frac{1}{2}$ and $y = 1$ have nonanalytic holonomy, while all the other leaves have trivial holonomy. The leaves of the graph lying over the compact leaves with trivial holonomy are tori, the leaves lying over the noncompact leaves are planes, and the leaves lying over the leaves with nontrivial holonomy are cylinders. These latter leaves may be thought of as winding around the tori so as to form a covering, but points in the same fiber cannot be separated, so the graph is not Hausdorff. The plane leaves wind toward the cylinder leaves, but in this case the winding is already visible in the original foliation.

2.47 Example. For the foliation of dimension 0 of M, the graph is the diagonal, and for the foliation of codimension 0, the graph is $M \times M$.

An important tool in the study of a locally compact group is the space of complex-valued functions which are integrable with respect to Haar measure on the group. These functions form an algebra under convolution, and there are an involution and a norm so that they form a C^*-algebra in the sense of the following definition.

2.48 Definition. A *C^*-algebra* is a complex Banach space $\mathscr{A}$ with a multiplication $(x, y) \mapsto xy$ and an involution $x \to x^*$ such that $\mathscr{A}$ is an algebra over the complex numbers satisfying:

$$
\begin{aligned}
(x^*)^* &= x \\
(x + y)^* &= x^* + y^* \\
(\lambda x)^* &= \bar{\lambda} x^* \\
(xy)^* &= y^* x^* \\
\|xy\| &\leqq \|x\| \, \|y\| \\
\|x^*\| &= \|x\| \\
\|x^* x\| &= \|x\|^2.
\end{aligned}
$$

A C^*-algebra is a very rich structure, with which one can associate many other structures by means of algebraic constructions. For example, by considering modules over the algebra, one can obtain a K-theory, even when the algebra is not associated with any manageable topological space. Thus, it is useful to try to associate a C^*-algebra to the holonomy groupoid of a foliation, so as to generalize the algebra of a locally compact group. In general, there is no natural measure on a groupoid, so instead of using functions to construct the algebra, one must use measures. Since the groupoid in question is a manifold, these measures are best obtained by using the differential geometric idea of "density".

2.49 Definition. Given a p-dimensional vector bundle, the associated *bundle of densities of order* α is the 1-dimensional vector bundle obtained from the homomorphism h of GL_p into GL_1 given by

$$h(A) = |\det(A)|^\alpha.$$

A *density* is a section of a bundle of densities.

This definition makes sense for real or complex vector bundles and for all real α, giving a bundle whose group is the multiplicative group of positive reals. Since this group takes a positive real number to a positive real number, the notion of positive density makes sense. If $\alpha = 1$ and the original bundle is the tangent bundle of a p-dimensional manifold, any positive density defines a positive measure. A p-form on a p-dimensional manifold differs from a density of order 1 only in that the absolute value is omitted from the formula defining h. Since the order of factors is irrelevant for Fubini's theorem, this theorem deals with densities or measures, not with p-forms.

To define the C^*-algebra associated with the holonomy groupoid of a C^∞ foliated manifold $(M,\mathscr{F})$, let Ω_M^α be the bundle of densities of order α associated with the tangent bundle along the leaves, and let

$$\Omega_\Gamma^{1/2} = (Z \oplus A)^* (\Omega_M^{1/2} \otimes \Omega_M^{1/2}).$$

Let $C_c^\infty(\Omega_\Gamma^{1/2})$ be the space of compactly supported C^∞ sections of $\Omega_\Gamma^{1/2}$, let $k_i \in C_c^\infty(\Omega_\Gamma^{1/2})$, and for $\gamma \in \Gamma(\mathscr{F})$, define

$$(k_1 * k_2)(\gamma) = \int_{\gamma_1 \gamma_2 = \gamma} k_1(\gamma_1) k_2(\gamma_2)$$

$$k_1^*(\gamma) = \overline{k_1(\gamma^{-1})}.$$

For every $x \in M$, let

$$\Gamma_x = \{\gamma \mid A(\gamma) = x\}$$

$$(\pi_x(k_1)\xi)(\gamma) = \int_{\gamma_1 \gamma_2 = \gamma} k_1(\gamma_1)\xi(\gamma_2)$$

where ξ is a square integrable density of order $\frac{1}{2}$ on Γ_x. The norm is defined by

$$|k| = \sup_{x \in M} |\pi_x(k)|$$

where $|\pi_x(k)|$ is the norm of $\pi_x(k)$ considered as an operator in the Hilbert space of square integrable densities of order $\frac{1}{2}$ on Γ_x.

2.50 Definition. The *C^*-algebra of* $(M,\mathscr{F})$ is the completion of $C_c^\infty(\Omega_\Gamma^{1/2})$ with respect to the norm defined above.

Suppose now that the manifold is compact and the foliation admits a non-negative transversal invariant measure Λ of finite mass. Then by Theorem 2.8, there is a corresponding foliation cycle of order 0, defining an element of $H_p(M;\mathbb{R})$. Evaluating the Euler class of the tangent bundle along the leaves on this foliation cycle gives a number which may be described as the Euler number of the foliation. It is possible to give an alternative description of this Euler number as the alternating sum of the average Betti numbers of the holonomy coverings of the leaves of $\mathscr{F}$. There are two difficulties in doing this: the Betti numbers of the individual leaves may be infinite, and the integration needs to be performed over a very bad space, the space of leaves. To deal with these problems, Connes [1979] has introduced a theory of integration which desingularizes the space of leaves. To apply it to the problem at hand, one takes as the cohomology space to be considered the space $h^j(\tilde{f})$ of square integrable forms on the holonomy covering $\tilde{f}$ which are harmonic with respect to a Riemannian structure along the leaves.

2.51 Theorem. *The scalar $\beta_j = \int \dim h^j(\tilde{f})\, d\Lambda(f)$ is finite and independent of the choice of Riemannian structure. Moreover, $\sum(-1)^j \beta_j$ is equal to the Euler number of the foliation.*

Proof. See Connes [1979]. □

The transverse invariant measure, combined with the theory of integration over singular spaces, also gives a trace on the C^*-algebra of $\mathscr{F}$. This makes the algebra much more manageable. The above theorem is only a sample of the known results, but much remains unknown, and the theory is not yet in definitive form. For a survey of its state, see the papers of Connes [1980, 1981].

The graph of a foliation was introduced by Thom [1964]. It was later studied in some detail by Winkelnkemper [1982], who was interested primarily in Riemannian foliations (see § 4 below) and geometric properties of the graph. The study of its analytic properties was started by Connes [1979] and seems likely to continue for some time in the future.

3. Foliations of a Riemannian Manifold

Since any smooth manifold admits (positive definite) Riemannian metrics, it is natural to make use of such a metric to get information about any other structure which may be given on the manifold. There are two difficulties in this approach: the conclusions may depend on the choice of metric, and there may be no metric which is nicely related to the structure. For a foliation of codimension q, there also exist metrics with q negative eigenvalues and the rest positive, but these metrics may also fail to be nicely adapted to the foliation in other respects. Since any such metric is canonically associated to a positive definite metric obtained by changing the sign of the metric induced on the plane field orthogonal to the foliation, and since indefinite metrics are harder to work with, there may be little profit in considering the indefinite metrics. However, there is at least one theory where the indefinite metric is useful, namely, the theory of general relativity. Here the fundamental object of study is a manifold with an indefinite metric, but it has proved useful in solving certain problems to suppose (or prove) the existence of a space-like foliation of codimension 1 as an additional tool. Since the leaves of any foliation are immersed submanifolds, the Riemannian theory of foliations is closely related to the classical theory of submanifolds of a (possibly indefinite) Riemannian manifold. The principal difference is that the first and second fundamental form become tensors defined on the whole manifold, and that their analogs for the normal plane field are also needed for the study.

Before considering foliated Riemannian manifolds in more detail, it will be convenient to introduce some concepts which are defined for any C^k manifold M, supplied with a C^{k-1} positive definite Riemannian metric, $k \geqq 1$, such that M is complete with respect to the corresponding topological metric.

3.1 Definition. The *growth function* of a connected, complete, positive definite Riemannian manifold M is the function G defined by 2.22 with respect to its natural foliation of codimension 0. The notions of *exponential* and *polynomial growth* of M are defined by use of G.

It is easy to see that the growth type does not depend on the choice of base point.

On a compact, orientable n-dimensional manifold, the exterior derivative of every smooth $(n-1)$ form must equal 0 at some point, since otherwise its integral over the manifold is both 0 and not equal to 0. Also, if the manifold is Riemannian, every form is bounded, in the sense that its norm with respect to the metric is bounded. For complete manifolds which are not compact, an important distinction arises in the attempt to generalize these facts (compare Definition 2.23).

3.2 Definition. A complete positive definite n-dimensional Riemannian manifold is said to be *closed at infinity* if for every bounded $(n-1)$-form η, $d\eta$ is either unbounded or not bounded away from 0.

For example, a simply connected complete manifold of constant curvature is compact if the curvature is positive, closed at infinity if the curvature is 0, and not closed at infinity if the curvature is negative.

3.3 Proposition. *If M is not closed at infinity, then the volume of any compact region is bounded above by a constant times the volume of its boundary, and furthermore M has exponential growth. If M is compact, it has polynomial growth of degree zero.*

Proof. The first statement follows from Stokes' theorem. If M is compact, the growth function is bounded by the volume of M. □

The converse of the second statement is false even if M is complete since there exist closed unbounded subsets of $\mathbb{R}^3$ which are smooth 2-manifolds and have finite 2-dimensional volume. The precise relation between closure at infinity and growth is not understood. Further properties of closure at infinity can be found in the papers of Sullivan [1976a] and Sullivan and Williams [1976]. Clearly, the behavior at infinity is preserved by isometries, but there is a much larger class of diffeomorphisms for which it is preserved.

3.4 Definition. A C^1 diffeomorphism $f\colon M_1 \to M_2$ of manifolds with positive definite Riemannian metrics g_i, $i = 1, 2$, is a *quasi-isometry* if there exist positive constants e and E such that for any pair of tangent vectors (X, Y) to M_1,

$$e g_1(X, Y) \leqq g_2(f_* X, f_* Y) \subseteq E g_1(X, Y).$$

M_1 and M_2 are *quasi-isometric* if there exists a quasi-isometry from M_1 to M_2.

3.5 Proposition. *The relation of being quasi-isometric is an equivalence relation. Any quasi-isometry f can be factored as a composition $f_1 \circ f_2$, where f_1 is an isometry and f_2, considered as a mapping of the underlying manifolds, is the identity. Every C^1 diffeomorphism of compact manifolds is a quasi-isometry. Quasi-isometries preserve completeness, exponential growth, nonexponential growth, polynomial growth of degree d, and closure at infinity.*

Proof. The first statement is obvious. Let $f\colon (M_1, g_1) \to (M_2, g_2)$ be a quasi-isometry. Take on M_1 the metric $(T^* f) g_2$, and let f_1 be f considered as a mapping from $(M_1, (T_* f) g_2)$ to (M_2, g_2), and f_2 be the identity mapping of M_1. The minimum eigenvalue $e(x)$ of $(T_* f) g_2$ with respect to g_1 is positive and varies continuously with x, hence has a positive minimum e if M_1 is compact. Likewise, the largest eigenvalues have a finite maximum E, and these two values satisfy the conditions of the definition of quasi-isometry. Since any isometry preserves all the listed properties, it is sufficient to consider the case that f is the identity map of the underlying differentiable manifold $M = M_1 = M_2$. Furthermore, any quasi-isometry takes a Cauchy sequence to a Cauchy sequence and a form bounded away from 0 or infinity to another with the same property, so only the growth types remain to be considered. Let $D_i(x, r)$ denote the set of points whose distance from x with respect to g_i is less than r, and let $V_{i,j}(x, r)$ the volume with respect to g_i of $D_j(x, r)$. Suppose M has exponential growth with respect to g_1, that is,

$$\liminf_{r \to \infty} \left\{ \frac{1}{r} \log V_{1,1}(x, r) \right\} > 0.$$

Since the identity is a quasi-isometry

$$D_1(x, er) \subseteq D_2(x, r) \subseteq D_1(x, Er),$$

so

$$V_{1,1}(x, er) \leqq V_{1,2}(x, r) \leqq V_{1,1}(x, Er).$$

Hence

$$V_{2,2}(x, r) \geqq e^n V_{1,2}(x, r) \geqq e^n V_{1,1}(x, er),$$

so that for r sufficiently large

$$\log V_{2,2}(x, r) \geqq n \log e + \log V_{1,1}(x, er) \geqq kr$$

for some constant k. It follows that M has exponential growth with respect to g_2. By symmetry, if M has nonexponential growth with respect to g_1, it has also with respect to g_2. Finally, if

$$V_{1,1}(x, r) \leqq P(r)$$

where P is a polynomial of degree d, then

$$V_{2,2}(x, r) \leqq E^n V_{1,2}(x, r) \leqq E^n V_{1,1}(x, Er) \leqq E^n P(Er)$$

where the last expression is also a polynomial of degree d in r. This completes the proof of the proposition. □

Suppose now that the manifold M also has a C^k foliation $\mathscr{F}$ of codimension q. Then each leaf has an induced Riemannian metric, and it is desirable to find properties of the induced metric which do not depend on the choice of metric on M.

3.6 Proposition. *If M is complete, then each leaf is complete with respect to the induced metric. If M is compact, then:*

(i) *The identity map is a quasi-isometry between any two metries induced on a leaf by metrics on M. In particular, growth type and closure at infinity for a leaf do not depend on the choice of metric for M.*

(ii) *Any leaf with polynomial growth of degree* 0 *(that is, finite volume) is compact.*

(iii) *If some leaf is closed at infinity, then there exists a transversal invariant measure with support in its closure.*

Proof. A Cauchy sequence $\{x_n\}$ on a leaf L is necessarily Cauchy in M, hence converges to a point x of M. Consider a pair of compact flat neighborhoods of x, with the smaller K contained in the interior of the larger K_1. For each plaque P of K_1 which meets K, there is a positive minimum distance $r(P)$ on its leaf between $P \cap K$ and the complement of K_1, and there is a positive number r less than half the minimum over all P of the numbers $r(P)$. Then for any leaf L, any two points of $L \cap K$ whose distance from each other in the metric of the leaf is less than r must be in the same plaque. In particular, almost all of the Cauchy sequence $\{x_n\}$ lies in a single plaque, so that x lies in that plaque and therefore in L.

Suppose M is compact, and two metrics g_1 and g_2 are given. It suffices to consider the positive definite version of these metrics. Then each leaf is complete and also the constants e and E can be defined, with respect to the identity map of M as in the proof of Proposition 3.5. The inequalities involving e and E also apply to the restriction to any leaf, so (i) holds. (ii) has already been proved as part of Proposition 2.29. An alternative proof can be based on the observation that there is a lower bound to the volume of a plaque, and a noncompact leaf necessarily contains infinitely many disjoint plaques. (iii) is proved by Sullivan [1976a] as a consequence of his discussion of forms positive on leaves (see Proposition 2.10). □

One consequence of Proposition 3.6 is that closure at infinity and nonexponential growth have more or less the same meaning for a leaf of a foliation of a compact manifold. This adds interest to the question of whether they are equivalent intrinsically, but does not resolve it, since even polynomial growth of degree 0 has stronger implications for a leaf than it does in general.

Thus far in the discussion of foliated Riemannian manifolds, only the metric induced on the leaves has been considered. However, the most interesting parts of submanifold theory involve connections and curvature. These parts will now be reviewed in a form suitable for later applications. Given a manifold M and a submanifold L, the covariant derivative of M can be used to define derivatives with respect to tangent vectors of L. If X and Y are both tangent vector fields, then the tangent component of $\nabla_X Y$ is the same as the covariant derivative with respect to the metric induced on L. If also N is a normal field, then a connection on the normal bundle is defined by taking as the derivative of N the normal component of $\nabla_X N$. The rest of the information about ∇_X is contained in the tensors K_N defined by

$$K_N(X, Y) = -\langle N, \nabla_X Y\rangle = \langle \nabla_X N, Y\rangle. \tag{3.7}$$

K_N is called the second fundamental form with respect to N. If X and Y are tangent vector fields belonging to an orthonormal basis, then

$$\langle N, \nabla_X Y\rangle = \langle N, \nabla_Y X\rangle + \langle N, [X, Y]\rangle.$$

Since the Lie bracket of tangent vectors is tangent, it follows that K_N is symmetric. If M has constant curvature, the induced metric (first fundamental form), the normal connection, and the tensors K_N determine the embedding of L up to an isometry of M. Thus, they contain all the geometric information.

It is sometimes convenient to replace the K_N by other tensors containing the same information but having different variance properties (raise one or both indices, in the index notation). Thus, one may consider the field of linear transformations (or the vector-valued 1-form) k_N defined by taking

$$\langle k_N(X), Y\rangle = K_N(X, Y) = \langle \nabla_X N, Y\rangle \tag{3.8}$$

and setting the normal component of $k_N(X)$ equal to 0. (k_N is sometimes called the Weingarten tensor because of its occurrence in the Weingarten formula giving the relation between the two connections.) Both K_N and k_N are tensors on the submanifold L. The domain of K_N may be extended to tangent vectors to M at points of L as follows: First, define a tensor K^N by

$$K^N(\phi, \psi) = K_N(X_\phi, X_\psi) = \psi(k_N(X_\phi)) \tag{3.9}$$

where (ϕ, ψ) is a pair of 1-forms and X_ϕ is the vector defined by

$$\zeta(X_\phi) = \langle \zeta, \phi\rangle$$

for all 1-forms ζ. Second, map K^N by using the injection map of L into M. The image tensor, also denoted K^N, vanishes whenever one argument is a form which vanishes on L. Third, change variance again, using the metric on M. The K_N thus obtained vanishes whenever one argument is orthogonal to L. Of course, the last statement could be used to define the extension of K_N, but the multistep argument is less arbitrary, and gives some indication of how K^N is used.

The second fundamental form is often called the extrinsic, or relative, curvature of L, because it measures the difference between the sectional curvature of a 2-plane tangent to L as viewed in L or in M. The formula which does this is called the Gauss equation. Of course, there are equivalent formulas expressing the difference between the values of the Riemann curvature tensors of L and M on a quadruple of vectors tangent to L. The Codazzi-Mainardi equation gives the value of the Riemann curvature tensor on a quadruple including exactly one normal vector in terms of the vector valued 2-form dk_N. Finally, the Ricci equation gives the value of the Riemann tensor on certain quadruples involving 2 normal vectors in terms of K_N and the curvature of the induced connection on the normal bundle. The preceding discussion assumes that L is at least two-dimensional. For a curve, the intrinsic curvature is 0, and the curvature referred to in the usual discussions is a special case of the second fundamental form. The rest of the coefficients which occur in Frenet-Serret formulas give a canonical form for the normal connection.

There are many other uses for the second fundamental form besides measuring the differences between connections and between curvatures. For example, a submanifold is said to be totally geodesic if its geodesics in the induced metric are also geodesics in the containing manifold. This is equivalent to the vanishing of K_N. Also, k_N is related to the classical Plateau problem of finding the submanifold of smallest volume with a given boundary. Indeed, a necessary condition that the volume be minimal is that the trace $\operatorname{tr}(k_N)$ be 0 for all N. In general, the first variation of volume is given by the mean curvature vector, a normal vector $\operatorname{tr}(k)$

whose component in the direction N is given by $\operatorname{tr}(k_N)$. For embeddings of L of codimension 1 in $\mathbb{R}^m$, the Gauss map is the map of L into the unit sphere that takes each point to the unit normal at that point. This map is related to $\det(k_n)$. K^N is convenient to use in problems in which L is the object of primary interest, such as the theory of deformations of Riemannian structures, or the initial value problem of general relativity. In the former, one considers fiber bundles with fiber L, and studies the changes in induced metric from fiber to fiber. In the latter, a positive definite metric and K^N are given on L, and it is desired to find an indefinite metric on M which satiesfies Einstein's equations and induces the given data on L.

After this review of the properties of submanifolds, it is time to turn to the discussion of foliations. Since every point lies on a leaf, the induced Riemannian metric and the forms K_N are tensors defined on open sets rather than on a submanifold. The connections induced on the tangent and normal bundles of a submanifold also fit together nicely, but the result is now only two partial connections in the sense that the bundles become bundles on M, but differentiation is defined only with respect to vectors tangent to a leaf. To extend these partial connections to full connections, one can repeat the constructions, interchanging the roles of tangent and normal. There are some difficulties in imitating the full theory, since the normal plane field may not be integrable, and its integrability tensor must be taken account of in the equations. One can, in fact, give a rather complete discussion in which the data are merely a pair of orthogonal complementary plane fields. Important formulas for many special cases are obtained as corollaries, and thereby put into a natural context.

Let A, B, C denote arbitrary vector fields on a manifold with a Riemannian metric $\langle\, ,\rangle$. Let P be a field of linear transformations satisfying

$$P^2 = P \qquad \langle PA, B\rangle = \langle A, PB\rangle. \tag{3.10}$$

Then, if I is the identity transformation, and $Q = I - P$, the tangent bundle is the orthogonal direct sum of subbundles which are images of the projections P and Q, and which will be known as the vertical and horizontal spaces respectively. Some further identities are:

$$Q^2 = Q \qquad \langle QA, B\rangle = \langle A, QB\rangle$$
$$P + Q = I \qquad PQ = QP = 0.$$

Let L, M, N be sections of the image of Q and X, Y, Z be sections of the image of P. Then the difference of P and Q is a field of linear transformations G which satisfies

$$G^2 = I \qquad \langle GA, GB\rangle = \langle A, B\rangle. \tag{3.11}$$

Conversely, given G with these properties, P may be defined by

$$P = \tfrac{1}{2}(I + G)$$

and will have the properties (3.10). Indeed, P and Q project onto the eigenspaces of G for the eigenvalues $+1$ and -1 respectively. Some further useful formulas are

$$G = P - Q \qquad Q = \tfrac{1}{2}(I - G)$$
$$PG = GP = P \qquad QG = GQ = -Q$$

For present purposes, it is most convenient to work with the tensor G and with the form γ defined by

$$\gamma(A, B) = \langle GA, B \rangle \tag{3.12}$$

γ is symmetric and nonsingular, so it can be viewed as a metric for which the vectors X have positive length and the vectors N have negative length. For the moment, however, it will be viewed as a symmetric 2-form, to be analyzed with respect to the given Riemannian metric and its torsion-free connection ∇.

3.13 Definition. The *almost product connection* ∇^G associated to G is given by the formulas:

$$\nabla^G_A X = P\nabla_A X$$
$$\nabla^G_A N = Q\nabla_A N.$$

The *adapted connection* $\nabla^{\mathcal{T}}$ associated to G is given by:

$$\nabla^{\mathcal{T}}_Y X = P\nabla_Y X$$
$$\nabla^{\mathcal{T}}_N X = P[N, X]$$
$$\nabla^{\mathcal{T}}_Y M = Q[Y, M]$$
$$\nabla^{\mathcal{T}}_N M = Q\nabla_N M.$$

∇^G preserves the metric, but its torsion tensor T^G is not in general zero. T^G can be expressed in terms of the discarded components of $\nabla_A B$, but the formulas will not be given until the general discussion of tensors arising from these components has been completed. This discussion is best carried out in terms of the covariant derivatives of γ. Let

$$\nabla\gamma(C, A, B) = (\nabla_C \gamma)(A, B). \tag{3.14}$$

3.15 Lemma. $\nabla\gamma$ *satisfies the following identities:*

$$(\nabla\gamma)(C, A, B) = \langle (\nabla_C G) A, B \rangle$$
$$(\nabla\gamma)(C, A, B) = (\nabla\gamma)(C, B, A)$$
$$(\nabla\gamma)(C, X, Y) = 0$$
$$(\nabla\gamma)(C, M, N) = 0.$$

Proof. The first identity follows from the fact that ∇ preserves the metric, and the second from the fact that γ is symmetric. Since $G^2 = I$,

$$(\nabla G) G + G(\nabla G) = 0$$

which leads to the identity

$$\langle (\nabla_C G) GA, GB \rangle + \langle (\nabla_C G) A, B \rangle = 0.$$

Because of the symmetry of $\nabla\gamma$ in its last two variables, this formula is equivalent to the last two identities of Lemma 3.15. The statement and proof of this lemma are due to Naveira [1982]. □

The group which preserves both G and the metric at any point is the product $O(p)\times O(q)$ of orthogonal groups. Thus, it is natural to decompose $\nabla\gamma$ into components lying in the irreducible invariant subspaces of the tensor space $T^*\otimes T^*\otimes T^*$ under the action of this group. Let P^* (respectively Q^*) denote the subspace of T^* consisting of forms which vanish on Q (respectively P). Then $\nabla\gamma$ lies in

$$P^*\otimes P^*\otimes Q^* + Q^*\otimes Q^*\otimes P^*$$

and in each term the third component is independent of the first two. Furthemore $P^*\otimes P^*$ is decomposed into antisymmetric and symmetric parts, with the latter decomposed into a trace free component and a multiple of the metric. A similar decomposition holds for $Q^*\otimes Q^*$. Thus, there are six terms in the decomposition of $\nabla\gamma$.

3.16 Definition. The *second fundamental form* of the horizontal plane field is the tensor $\mathscr{S}$ defined by

$$\begin{aligned}\langle \mathscr{S}_M N, X\rangle &= \tfrac{1}{2}\langle \nabla_M N + \nabla_N M, X\rangle\\ \langle \mathscr{S}_M N, L\rangle &= 0\\ \langle \mathscr{S}_M X, Y\rangle &= 0\\ \langle \mathscr{S}_M X, N\rangle &= \langle \mathscr{S}_M N, X\rangle\\ \mathscr{S}_X &= 0.\end{aligned}$$

The second fundamental form $\mathscr{T}$ of the vertical plane field is defined by the same equations with horizontal and vertical vectors interchanged.

The name "second fundamental form" is appropriate, since if the horizontal field is integrable

$$\langle \mathscr{S}_M N, X\rangle = -K_X(M,N)$$

where K_X is the second fundamental form (3.7) of the leaves as submanifolds.

3.17 Proposition. *The antisymmetric parts of $\nabla\gamma$ are given by tensors $\mathscr{A}$ and $\mathscr{B}$. $\mathscr{A}$ contains the same information as the integrability tensor A_Q of the horizontal plane field and is defined by*

$$\langle \mathscr{A}_M N, X\rangle = \tfrac{1}{2}\langle \nabla_M N - \nabla_N M, X\rangle$$

with the rest of the equations as in the definition of $\mathscr{S}$. $\mathscr{B}$ contains the same information as the integrability tensor A_P of the vertical plane field, and is defined by the same equations as $\mathscr{A}$ with horizontal and vertical vectors interchanged. The symmetric parts of $\nabla\gamma$ are given by the second fundamental forms $\mathscr{S}$ and $\mathscr{T}$ of the horizontal and vertical plane fields respectively.

Proof. In the subspace $Q^* \otimes Q^* \otimes P^*$, the antisymmetric part $\mathscr{A}$ of $\nabla\gamma$ satisfies the identities

$$\begin{aligned}
\mathscr{A}(C,A,B) &= \mathscr{A}(C,B,A)\\
\mathscr{A}(C,X,Y) &= 0\\
\mathscr{A}(C,M,N) &= 0\\
\mathscr{A}(M,N,X) &= \tfrac{1}{2}\langle \nabla_M N - \nabla_N M, X\rangle\\
\mathscr{A}(X,Y,N) &= 0.
\end{aligned}$$

Letting $\mathscr{A}(C,A,B) = \langle \mathscr{A}_C A, B\rangle$ yields the above formulas for $\mathscr{A}_C$. Since

$$\nabla_M N - \nabla_N M = [M,N],$$

$\mathscr{A}_C$ vanishes if and only if the horizontal field is integrable. The discussion of $\mathscr{S}_C$ is similar, except that

$$\mathscr{S}(M,N,X) = \tfrac{1}{2}\langle \nabla_M N + \nabla_N M, X\rangle.$$

The proof is concluded by interchanging horizontal and vertical to derive $\mathscr{B}$ and $\mathscr{T}$. □

According to the general principle mentioned above, each second fundamental form admits a further decomposition into a form of trace 0 and a multiple of the metric. Let $\{X_a, N_\alpha\}$ be an orthonormal basis adapted to the two plane fields.

3.18 Proposition. *The second fundamental form $\mathscr{S}$ decomposes into a form which contains the same information as the mean curvature vector of the plane field and a trace free form. The mean curvature vector is the normal vector to the horizontal field whose component in the X direction is given by*

$$\sum_{\alpha=1}^{q} \langle \nabla_{N_\alpha} N_\alpha, X\rangle$$

while the corresponding component $\mathscr{S}^\perp$ of $\mathscr{S}$ is given by

$$\mathscr{S}^\perp(L,M,X) = \frac{1}{q}\left\{\sum_{\alpha=1}^{q} \langle \nabla_{N_\alpha} N_\alpha, X\rangle\right\}\langle L,M\rangle$$

which the rest of the defining equations unchanged.

The trace-free component $\mathscr{S}^\top$ of $\mathscr{S}$ is given by

$$\mathscr{S}^\top(L,M,X) = \mathscr{S}(L,M,X) - \mathscr{S}^\perp(L,M,X),$$

and is characterized by the property that for all X,

$$\sum_{\alpha=1}^{q} \mathscr{S}(N_\alpha, N_\alpha, X) = 0.$$

The formulas for $\mathscr{T}^\top$ and $\mathscr{T}^\perp$ are obtained by interchanging horizontal and vertical vectors.

So far, the use of the name "second fundamental form" for $\mathscr{S}$ and $\mathscr{T}$ has been justified solely on algebraic grounds. However, it also has some of the geometric properties of the usual second fundamental form.

3.19 Definition. A plane field on a Riemannian manifold is *totally geodesic* if each geodesic which is tangent to it at one point remains tangent for its entire length.

3.20 Proposition. *If a plane field is totally geodesic, then its second fundamental form vanishes. If the second fundamental form vanishes, then for every pair (X, Y) of tangent vector fields which are infinitesimal automorphisms of its normal field and for every normal vector field N*

$$N \cdot \langle X, Y\rangle = 0. \tag{3.21}$$

If in addition the normal field is integrable, then the given field is totally geodesic.

Proof. To use the established terminology, suppose the given field is vertical, so its second fundamental form is $\mathscr{T}$. Suppose the field is totally geodesic. Then any unit tangent vector can be extended locally to a unit vector field that is tangent both to a family of geodesics and to the plane field. Hence

$$\langle \mathscr{T}_X X, N\rangle = \langle \nabla_X X, N\rangle = 0$$

for all tangent vectors X and normal vectors N. Since

$$\langle \mathscr{T}_X Y, N\rangle = \langle \mathscr{T}_Y X, N\rangle,$$

the preceding equation implies $\langle \mathscr{T}_X Y, N\rangle = 0$, and therefore $\mathscr{T} = 0$.

Conversely, suppose $\mathscr{T} = 0$. Since the connection is Riemannian

$$\begin{aligned} N \cdot \langle X, Y\rangle &= \langle \nabla_N X, Y\rangle + \langle X, \nabla_N Y\rangle \\ &= \langle \nabla_X N, Y\rangle + \langle X, \nabla_Y N\rangle + \langle [N, X], Y\rangle + \langle X, [N, Y]\rangle. \end{aligned}$$

If X and Y are infinitesimal automorphisms of the horizontal field, then the last two terms vanish. On the other hand, the equation

$$\langle \mathscr{T}_X Y, N\rangle = -\tfrac{1}{2}\{\langle \nabla_X N, Y\rangle + \langle X, \nabla_Y N\rangle\}$$

implies

$$N \cdot \langle X, Y\rangle = -2\langle \mathscr{T}_Y Y, N\rangle = 0.$$

In general, there may not be many vector fields X which are infinitesimal automorphisms of the horizontal plane field. However, if the horizontal field is integrable, then locally it has many infinitesimal automorphisms. Indeed, if the field is defined by $dx^a = 0$, then every vector field of the form

$$A^a(x)\frac{\partial}{\partial x^a} + B^\alpha(x, y)\frac{\partial}{\partial y^\alpha}$$

is an infinitesimal automorphism. Furthermore, equation (3.21) implies that there is a metric on the local quotient such that the local projection is a Riemannian submersion. Hence, any vertical curve projects horizontally onto a curve of the same length, so that a vertical curve which projects onto a geodesic is a geodesic. Hence, the vertical field is totally geodesic, as required. This completes the proof. □

The notion of a totally geodesic plane field is of considerable geometric interest. The next section will be devoted to the study of Riemannian foliations, which on a Riemannian manifold can be characterized by the condition that the normal plane field be totally geodesic. An instructive example is a foliation whose leaves are the fibers of a locally trivial fibration. The metric on the total space is obtained by pulling back a metric from the base and adding an arbitrary metric along the fibers. (See Example 4.7.) If the total space is compact, then this metric can be chosen so that the fibers are also totally geodesic if and only if the structure group of the bundle can be reduced to a compact Lie group (Hermann [1960]).

In general, it is difficult to determine when a foliated manifold admits a metric such that the foliation is totally geodesic. In codimension 1, a necessary condition is that the Pontrjagin ring of the manifold vanish in the top dimension. This is proved by Johnson and Naveira [1981], using a connection on the tangent bundle to the foliation. The other known results deal with foliations of dimension 1. A foliation of a compact manifold by circles can be made geodesic if and only if the volume of the leaves is bounded (Wadsley [1975]). The suspension of a diffeomorphism is a flow which can be made geodesic, and suspensions are the only foliations of a compact surface which can be made geodesic (Gluck [1980]). Moreover, the only Morse-Smale flows which can be made geodesic are suspensions (Asimov and Gluck [1980]). A useful tool is a theorem of Sullivan [1978] characterizing geodesible flows in terms of foliation cycles. Alternatively, one can fix a Riemannian manifold and ask whether it admits any foliations by geodesics. In this direction, foliations of S^3 by great circles have been classified (Gluck and Warner [1983]). On the other hand, it is known that in general manifolds of positive curvature admit very few totally geodesic foliations in any dimension (Abe [1971], Ferus [1970], Dombrowski [1978]).

One of the problems in studying totally geodesic foliations of higher dimensions is that the natural generalization of Sullivan's result is not to totally geodesic foliations, but to minimal foliations, which will now be introduced.

3.22 Definition. A plane field is *minimal* if the trace of its second fundamental form $\mathscr{S}$ is 0. It is *umbilical* if $\mathscr{S} = \mathscr{S}^{\perp}$. A foliation is minimal or umbilical if its tangent plane field has the corresponding property.

Geometrically, an umbilical submanifold can be though of as one conformally equivalent to a totally geodesic submanifold. Thus, a foliation whose normal plane field is umbilical is locally conformally equivalent to a Riemannian foliation (Montesinos [1983]). Such a foliation is called a conformal foliation.

A submanifold is minimal in the sense used here if locally its volume is stationary under normal perturbations. This terminology is in accord with the ordinary usage of differential geometry, even though it is analogous to defining a minimum point of a function to be a place where all the first derivatives vanish.

The torsion T^G of ∇^G can also be described in terms of the decomposition of $\nabla\gamma$. In fact, the following formulas hold:

$$\langle T^G(X,Y),N\rangle = -\langle \mathscr{B}_X Y,N\rangle$$
$$\langle T^G(X,N),Y\rangle = \langle (\mathscr{B}_X + \mathscr{T}_X)Y,N\rangle$$
$$\langle T^G(X,N),M\rangle = \langle (\mathscr{A}_N + \mathscr{S}_N)M,X\rangle$$

$$\langle T^G(M,N),X\rangle = -\langle \mathscr{A}_M,N,X\rangle$$
$$T^G(N,X) = -T^G(X,N)$$
$$\langle T^G(X,Y),Z\rangle = \langle T^G(M,N),L\rangle = 0.$$

Of course, a Riemannian almost product structure admits other connections, which will be discussed as needed. One example is obtained by applying the construction of Walker (Proposition II.5.13) to get a connection, so that the torsion of the new connection depends only on $\mathscr{A}$ and $\mathscr{B}$. Another example is the torsion free connection associated to the indefinite metric γ.

In the case of a transversally oriented foliation of codimension 1, the unit normal N is unique, so the second fundamental form K_N of the leaves can be regarded as a quadratic form on the manifold M. Its determinant, trace, and the rest of the symmetric functions of the eigenvalues are real valued functions on M. Since the normal is one-dimensional, it is integrable, and the curvature of the normal curves can also be regarded as a real valued function on M. Thus, the integral of any of these functions over M with respect to the volume element is an invariant of the foliation. In some cases, this can be computed, and turns out to be independent of the foliation, depending only on the curvature and volume of M.

3.23 Proposition. *Let M be a compact C^{k+1} manifold, $k \geqq 1$, of dimension $m+1$ with a C^{k+1} transversally oriented foliation of codimension 1 and a C^k Riemannian metric of constant sectional curvature C. Let σ_i be the i-th elementary symmetric function of the eigenvalues of the second fundamental form of the leaves, and let dV be the volume element. Then, if i and m are even*

$$\int_M \sigma_i \, dV = \binom{m/2}{i/2} C^{i/2} \operatorname{vol} M$$

and in all other cases, the integral is 0.

Proof. For the case $i = m$, see Asimov [1978], and for the more general case, see Brito, Langevin, and Rosenberg [1977]. The latter authors produce explicitly a form whose exterior derivative is

$$\left\{\sigma_i - \binom{m/2}{i/2} C^{i/2}\right\} dV.$$

Since M has Euler number 0, the Gauss-Bonnet theorem implies that for m odd, $C = 0$. □

Given an isotopy class of foliations on a Riemannian manifold one can ask whether there is in the class a foliation which minimizes

$$\int_M |\sigma_i| \, dV.$$

Preliminary results of Langevin [1979] show that there may not be any such foliation. One may also consider the foliation as fixed, and look for a metric minimizing the integral. If $i < m$, very little is known, but if $i = m$, one seeks a metric making the given foliation minimal. Much is known about this problem in any dimension and codimension.

3.24 Proposition. *For an oriented foliation of a compact manifold, the following statements are equivalent:*

(i) *There is a metric for which the leaves are minimal submanifolds.*
(ii) *No foliation cycle is approximately the boundary of a tangent $(p+1)$-chain.*
(iii) *There is a relatively closed p-form which is positive on the leaves.*
(iv) *There is a smooth, positive, compactly supported, holonomy invariant function f on the transverse manifold, such that $df = 0$ in $\mathscr{D}_1(M/G)$ and f is strictly positive on a set meeting each leaf.*

If in addition the leaves are all compact, then the following statements are equivalent to the above:

(v) *The volume of the leaves is bounded.*
(vi) *Each leaf admits a fundamental system of saturated neighborhoods.*

If a metric along the leaves is already given, then it can be extended to a metric such that the leaves are minimal if and only if the associated volume along the leaves is the restriction of a relatively closed p-form.

Proof. The equivalence of the first three statements is proved by Sullivan [1979], while Haefliger [1980] shows that the fourth is also equivalent. (The notation in (iv) is that of Proposition 1.5.) The results on the compact case and the case of a preassigned metric along the leaves are due to Rummler [1979]. □

In particular, if there is an immersed compact cross-section, then there is a metric such that the leaves are minimal. This statement is the proper genealization of the fact that a flow with section is geodesible, just as the equivalence of (i) and (ii) in Proposition 3.24 generalizes a criterion for geodesible flows. By statement (v), the examples of compact foliations with bounded volume are also examples in which the leaves can be made minimal. Some foliations which cannot be made minimal are a compact foliation with unbounded volume, a Reeb component, and a horocycle foliation of the tangent sphere bundle of a compact manifold of negative curvature. Further examples are given by Haefliger, Rummler and Sullivan in the papers cited in the proof.

Minimal foliations can also be related to harmonic forms, but the appropriate notion of harmonic must first be introduced. The appropriate forms have values in the normal bundle of the foliation, and it is necessary to introduce an adapted connection (Definition II.5.18) in this bundle. The precise connection ${}_q\nabla$ needed is obtained by applying the construction given in the proof of Proposition II.5.21, using ∇^G for ∇'. ${}_q\nabla$ is therefore given by the formulas

$$\begin{cases} {}_q\nabla_X N = Q[X,N] \\ {}_q\nabla_M N = Q\nabla_M N \end{cases} \tag{3.25}$$

where ∇ is the Riemannian connection. The torsion ${}_qT$ of ${}_q\nabla$ vanishes (Definition II.5.10). A differential operator ${}_qd$ operating on forms with values in the normal bundle is defined by

$$({}_q d\omega)(A_1, \ldots, A_{r+1}) = \sum(-1)^{i+1} (\hat{\nabla}_{A_i} \omega)(A_1, \ldots, \hat{A}_i, \ldots, A_{r+1})$$

where

$$(\hat{\nabla}_A \omega)(A_1, \ldots, A_r) = {}_q\nabla_A \omega(A_1, \ldots, A_r) - \sum_{i=1}^{c} \omega(A_1, \ldots, \nabla_A A_i, \ldots, A_r).$$

Also ${}_q d^*$ is defined by

$${}_q d^* \omega = (-1)^{n(r+1)+1} * {}_q d * \omega$$

where a is an r-form and $*$ is the extension to vector valued forms of the usual $*$ operator on a Riemannian manifold.

3.26 Proposition. *If Q is viewed as a vector-valued 1-form, then*

$${}_q dQ = 0 \qquad {}_q d^* Q = 0$$

if and only if all the leaves of the foliation are minimal. The first equation always holds, but in general ${}_q d^* Q$ is the mean curvature vector of the foliation.

Proof. See the paper of Kamber and Tondeur [1982]. A form which satisfies this pair of equations is called "harmonic". If the foliation is Riemannian, stronger results can be obtained. These will be given in the next section. ▯

The characteristic classes for a foliated manifold are defined in a highly algebraic way, which reveals little of their geometric content. However, there are more informative formulas involving the tensors $\mathcal{S}$, $\mathcal{T}$, and $\mathcal{A}$ defined in this section. The characteristic classes are associated with the normal bundle, and can be defined by use of any adapted connection. For present purposes, the best choice is the connection ${}_q\nabla$ given by (3.25). The classes are then represented by differential forms which are obtained as products of certain forms h_i and C_i on M, $i \leqq q$, where h_i is defined only if i is odd (see Definition II.5.23). C_i is a form of degree $2i$, and h_i is a form of degree $2i - 1$ whose restriction to any leaf is closed, hence defines a cohomology class of the leaf. As might be expected, it is the Riemannian properties of the normal plane field which play the largest role in the formulas for the characteristic classes.

3.27 Proposition. *The vector field corresponding to h_1 by means of the metric is the mean curvature vector of the normal plane field, and in general, the restictions to the leaves of the h_i depend only on the second fundamental form of the normal plane field. The Godbillon-Vey class $h_1 c_1^q$ depends on the mean curvature vector and the integrability tensor of the normal plane field, and on the second fundamental form of the leaves. If the normal plane field is totally geodesic, all the leaf and foliation classes vanish.*

Proof. The proof is a straightforward calculation based on known formulas for h_i and C_i (see Reinhart [1977]). For the last statement, Proposition 3.20 is used. ▯

In the case of a foliation with trivial normal bundle, the h_i are defined for all i, and a different set of classes is produced. However, a specific choice of framing is needed in the definition.

3.28 Proposition. *The leaf class h_{2i} depends on the framing, the integrability tensor of the normal plane field, and the second fundamental form of the normal plane field. If the normal field admits a Riemannian parallel framing, the leaf class h_{2i} is independent of the choice among such framings, and if in addition the normal field is integrable, the leaf class is 0. In any case, the reduction modulo the integers of h_{2i} does not depend on the choice of framing.*

Proof. See Reinhart [1977]. The last statement depends upon identifying the class in question with the restriction to the leaf of the Pontrjagin character defined by Chern and Simons [1974]. □

This completes the discussion of foliations of Riemannian manifolds in general. In the next section, the special case in which the normal plane field is totally geodesic will be considered.

4. Riemannian Foliations

The preceding section dealt with plane fields on a Riemannian manifold. General properties were derived, and various geometrically natural special classes were studied. This section contains a detailed study of one class that is particularly important.

4.1 Definition. A *Riemannian foliation* is a foliation such that the underlying Γ-structure is Riemannian. A Riemannian metric on a foliated manifold is *bundle-like* if the normal plane field is totally geodesic.

There are many statements equivalent to these definitions and many important examples of such foliations. On the other hand, there are rather strong structure theorems, leading to a good geometric understanding of the possibilities. Interesting properties of Riemannian submersion, harmonic maps, and harmonic integrals have generalizations to Riemannian foliations. The graph of a Riemannian foliation is Hausdorff.

There are restrictions on the characteristic classes of both tangent and normal bundles, which imply that the usual secondary classes vanish for a Riemannian foliation, but which allow the introduction of another type of secondary classes. All of these facts will be discussed in this section, beginning with the various equivalents to the definitions.

4.2 Proposition. *For a foliated Riemannian manifold, the following conditions are equivalent:*

(i) *The normal plane field is totally geodesic.*
(ii) *The second fundamental form of the normal plane field vanishes.*
(iii) *For every pair (M, N) of normal vector fields and every tangent vector field X,*

$$(\nabla_X^{\mathcal{T}} g)(M, N) = 0,$$

where g is the Riemannian metric.

(iv) *For every pair (M, N) of normal vector fields defined on some open set which are infinitesimal automorphisms of the foliation and every vector field X tangent to the foliation*

$$X \cdot \langle M, N \rangle = 0.$$

(v) *In any flat neighborhood with coordinates (x, y), the metric can be written*

$$g_{ab}(x, y)\, \theta^a \theta^b + g_{\alpha\beta}(y)\, dy^\alpha dy^\beta$$

where the foliation is defined by $\{dy^\alpha = 0\}$ and $\{\theta^a\}$ is any basis for the subspace orthogonal to that generated by $\{dy^\alpha\}$ in the cotangent space.

(vi) *Let W be any flat neighborhod and let $\pi: W \to B$ be the corresponding local projection. Then there is a Riemannian metric on B such that for every point $P \in W$, $T_P \pi$ is an isometry between the normal space to the leaf through P and the tangent space to B at $\pi(P)$.*

(vii) *The adapted connection on the normal bundle consistent with the metric is (torsion-free) Riemannian.*

(viii) *The holonomy group at some point of the adapted connection on the normal bundle consistent with the metric is metric preserving.*

If these conditions hold, then they induce a unique Riemannian Γ-structure. Conversely, given a foliation such that the induced Γ-structure contains a Riemannian Γ-structure, for each such Riemannian Γ-structure there is a Riemannian metric satisfying these conditions which induces it.

Proof. By Proposition 3.20, (i) is equivalent to (ii), and (ii) implies (iv). To show that (ii) is equivalent to (iii), first observe that

$$(\nabla_X^{\mathcal{T}} g)(M, N) = 0$$

is equivalent to

$$X \cdot \langle M, N \rangle = \langle \nabla_X^{\mathcal{T}} M, N \rangle + \langle M, \nabla_X^{\mathcal{T}} N \rangle.$$

Using the definition of $\nabla_X^{\mathcal{T}} M$ and the fact that ∇ is Riemannian, one obtains that the above identity is equivalent to

$$0 = \langle \nabla_M N, X \rangle + \langle \nabla_N M, X \rangle,$$

as required. To show that (iv) implies (v), assume (iv) and choose in the flat coordinate neighborhood a basis $\{\theta^a, dy^\alpha\}$ for the 1-forms as described in (v). Suppose first θ^a has the form

$$\theta^a = dx^a - \theta_\alpha^a(x, y)\, dy^\alpha.$$

Then the dual basis is $\{\partial/\partial x^a, N_\alpha\}$ where

$$N_\alpha = \theta_\alpha^a(x, y)\, \partial/\partial x^a + \partial/\partial y^\alpha$$

and $\langle N_\alpha, \partial/\partial x^a \rangle = 0$. The metric has the form

$$g_{ab}(x, y)\, \theta^a \theta^b + g_{\alpha\beta}(x, y)\, dy^\alpha dy^\beta$$

where

$$g_{\alpha\beta} = \langle N_\alpha, N_\beta \rangle$$
$$g_{ab} = \langle \partial/\partial x^a, \partial/\partial x^b \rangle.$$

Since each vector N_α is an infinitesimal automorphism, $X \cdot g_{\alpha\beta} = 0$ for every tangent field X, that is, $g_{\alpha\beta}$ is a function of y alone, as required. Since each $\{\theta^a\}$ which satisfies the hypotheses of (iv) can be written

$$\theta^a = h_b^a(x, y)\,[dx^b - A_\alpha^b(x, y)\,dy^\alpha\},$$

(iv) implies (v). If (v) holds, the metric on the local base given by

$$g_{\alpha\beta}(y)\,dy^\alpha\,dy^\beta$$

has the properties required by (vi). Suppose (vi) holds. Then any smooth curve transverse to the leaves projects to a curve which is no shorter, and which is the same length if and only if the curve is normal to the leaves. It follows that when a geodesic in the base is lifted to a normal curve, the lifted curve is a geodesic. Because the geodesic with a given initial vector is unique, the normal field is totally geodesic. Hence, (vi) implies (i). To complete the proof, it remains to study the relation between these six conditions and the adapted connection ${}_q\nabla$ on the normal bundle consistent with the metric. ${}_q\nabla$ is defined by formula (3.25). Since ${}_q\nabla$ is torsion free, it will be Riemannian if and only if for every vector A and all normal vector fields M and N

$$\langle {}_q\nabla_A M, N\rangle + \langle M, {}_q\nabla_A N\rangle = A \cdot \langle M, N\rangle.$$

If A is a normal vector L, this follows from ${}_q\nabla_L = \nabla_L$. If A is a tangent vector X, then

$$\begin{aligned}\langle {}_q\nabla_X M, N\rangle + \langle M, {}_q\nabla_X N\rangle &= \langle \nabla_X M - \nabla_M X, N\rangle + \langle M, \nabla_X N - \nabla_N X\rangle \\ &= X \cdot \langle M, N\rangle + \langle X, \nabla_M N + \nabla_N M\rangle.\end{aligned}$$

Hence, (vi) holds if and only if (ii) holds. (vii) is the integrated form of (vi). Hence, all seven conditions are equivalent.

Suppose they hold, and consider a coordinate covering by flat neighborhoods U_μ with coordinates (x_μ, y_μ). The coordinate changes are of the form

$$\begin{cases} x_\mu = F_{\mu\nu}(x_\nu, y_\nu) \\ y_\mu = G_{\mu\nu}(y_\nu) \end{cases}$$

and they preserve the Riemannian metric. Hence, the Γ-structure, which is given by $\{G_{\mu\nu}\}$, preserves the metrics on the local bases, that is, it is Riemannian. Suppose conversely that the Γ-structure contains a Riemannian Γ-structure. For each such Riemannian Γ-structure there is a coordinate covering $\{U_\mu\}$ and metrics $g_{\alpha\beta}(y)\,dy_\mu^\alpha\,dy_\mu^\beta$ on the local bases which are preserved by the mappings $G_{\mu\nu}$. Choose a transverse plane field, and construct a metric so that this field is normal and the local projection satisfies (vi). It i important to observe that this metric is not unique, but depends upon the choice of a transverse plane field and a metric along the leaves. This completes the proof of Proposition 4.2. □

As has been observed in Chapter III, the normal bundle of a foliation can be described in terms of the associated Γ-structure as the local pullback of the tangent bundle of the model. The pullbacks match up on the overlap of coordinate neighborhoods because the elements of Γ induce an isomorphism on the tangent bundle. In the case of a Riemannian Γ-structure, the pullback bundle has a metric and connection defined by the metric and connection on the model space. In fact, this turns out to be another description of the basic connection consistent with the metric of the total space.

4.3 Proposition. *The normal bundle to a Riemannian foliation admits a connection which is locally pulled back from the Riemannian connection on the model space. This pullback connection is basic and Riemannian. Given any metric consistent with the Riemannian Γ-structure of the foliation, the basic connection on the normal bundle consistent with this metric is the same as the pullback connection.*

Proof. Given a vector bundle E over N and a map $f: M \to N$, the pullback bundle E_f fits into a commutative diagram

$$\begin{array}{ccc} E_f & \xrightarrow{F} & E \\ \downarrow & & \downarrow \\ M & \xrightarrow{f} & N \end{array}$$

where F is an isomorphism on each fiber. Thus, any local basis $\{s_\alpha\}$ for the sections of E gives rise to a local basis $\{t_\alpha\}$ for the sections of E_f, where $t_\alpha = F^{-1} \circ s_\alpha$. Given any connection ∇ on E, the pullback connection ∇^f on E_f is defined by

$$\nabla_v^f (h^\alpha t_\alpha) = [(v \cdot h^\alpha) t_\alpha + h^\alpha F^{-1} (\nabla_{f_* v} s_\alpha)\},$$

where v is a tangent vector to M. In the present situation, f is a local projection for the foliation and E is the tangent bundle to the local quotient. Since the Riemannian connection is uniquely defined by the metric, it is preserved by isometries, and therefore the local pullbacks of the Riemannian connection fit together to form a connection on the normal bundle to the foliation. Given an orthonormal frame $\{N_\alpha\}$ on a local quotient of the foliation and a vector field X tangent to the foliation, the pullback vector fields N_α^* satisfy the conditions that $\nabla_X^f N_\alpha^* = 0$ and $[X, N_\alpha^*]$ be tangent. N_α^* can be identified with a normal vector field to the foliation, so that any normal vector field may be written $\Sigma h^\alpha N_\alpha^*$. Since

$$\nabla_X^f (h^\alpha N_\alpha^*) = (X \cdot h^\alpha) N_\alpha^*$$

and

$$\langle [X, h^\alpha N_\alpha^*], N_\beta^* \rangle = \langle (X \cdot h^\alpha) N_\alpha^*, N_\beta^* \rangle,$$

the pullback connection is basic. On the other hand, for any normal vector v

$$\begin{aligned} \langle \nabla_v^f (h^\alpha N_\alpha^*), k^\beta N_\beta^* \rangle &+ \langle h^\alpha N_\alpha^*, \nabla_v^f (k^\beta N_\beta^*) \rangle \\ &= (v \cdot h^\alpha) k^\beta \delta_{\alpha\beta} + h^\alpha (v \cdot k^\beta) \delta_{\alpha\beta} \\ &\quad + h^\alpha k^\beta \{\langle \nabla_{f_* v} N_\alpha, N_\beta \rangle + \langle N_\alpha, \nabla_{f_* v} N_\beta \rangle\} \\ &= (v \cdot h^\alpha) k^\alpha + h^\alpha (v \cdot k^\alpha) \\ &= v (h^\alpha k^\beta) \langle N_\alpha^*, N_\beta^* \rangle \\ &= v \cdot \langle h^\alpha N_\alpha^*, k^\beta N_\beta^* \rangle. \end{aligned}$$

Thus, ∇^f preserves the metric. For any pair (M^*, N^*) of pullback vector fields

$$\nabla^f_{M^*} N^* - \nabla^f_{N^*} M^* - [M^*, N^*]$$

maps into 0 under the map f_* so is a vector field tangent to the foliation. Thus, every pullback vector field satisfies

$$\langle \nabla^f_{M^*} \mathrm{N}^* - \nabla^f_{N^*} M^* - [M^*, N^*], L \rangle = 0$$

for every normal vector field L. It follows that every pair of normal vector fields has this property, so ∇^f is torsion free, and thus Riemannian.

Consider the basic connection ${}_q\nabla$ on the normal bundle consistent with the metric of the kind descriped in the statement. By proposition 4.2, this connection is Riemannian, besides being basic. Thus, both connections ∇^f and ${}_q\nabla$ satisfy the conditions

$$\langle \nabla_X M, N \rangle = \langle [X, M], N \rangle$$
$$\langle \nabla_L M, N \rangle + \langle M, \nabla_L N \rangle = L \cdot \langle M, N \rangle$$
$$\langle \nabla_M N, L \rangle + \langle \nabla_N M, L \rangle = \langle [M, N], L \rangle$$

for all tangent vector fields X and normal fields L, M, and N. The last two equations imply

$$\begin{aligned} 2\langle \nabla_M N, L \rangle = {} & M \cdot \langle N, L \rangle + N \cdot \langle M, L \rangle - L \cdot \langle M, N \rangle \\ & + \langle [M, N], L \rangle + \langle [L, M] \cdot N \rangle + \langle M, [L, N] \rangle. \end{aligned}$$

Thus, these three equations imply that the two connections are equal. This completes the proof of Proposition 4.3. □

4.4 Definition. The *canonical connection* for a Riemannian foliation is the connection on the normal bundle pulled back from the associated Riemannian Γ-structure.

The preceding proposition and definition are the result of looking at bundle-like metrics from the point of view of Γ-structures. Locally, they are also Riemannian submersions, and this point of view leads to further interesting results.

4.5 Definition. A mapping $f: M \to B$ of Riemannian manifolds is called a *Riemannnian submersion* if the tangent map at each point $x \in M$ is onto and induces an isometry between the normal plane at x to the level set of f and the tangent plane to B at $f(x)$.

Thus, condition (vi) of Proposition 4.2 asserts that each local projection is a Riemannian submersion, provided that a suitable metric is chosen on the local base. Thus, the known relations between the curvatures of M and B extend to a bundle-like metric and its local base. Let R and K denote the Riemann and sectional curvatures of the bundle-like metric, ${}_qR$ and ${}_qK$ denote those of the canonical connection, and ${}_pR$ and ${}_pK$ denote those of the leaves. Furthemore, let P_{AB} denote the 2-plane spanned by the vectors A and B.

4.6 Proposition. *Consider a foliated manifold with a bundle-like metric. Then* ${}_qR(X,Y)=0$, ${}_qR(X,N)=0$, *and* ${}_qR(M,N)$ *is equal to the curvature operator in the local base computed with respect to the projections of M and N. The curvature operators and sectional curvatures are related by:*

$$\begin{aligned}
\langle R(M,N)L,L'\rangle &= \langle {}_qR(M,N)L,L'\rangle - 2\langle \mathscr{A}_M N, \mathscr{A}_L L'\rangle \\
&\quad + \langle \mathscr{A}_N L, \mathscr{A}_M L'\rangle + \langle \mathscr{A}_L M, \mathscr{A}_N L'\rangle \\
\langle R(M,N)L,X\rangle &= \langle (\nabla_L \mathscr{A})_M N, X\rangle - \langle \mathscr{A}_M N, \mathscr{T}_X L\rangle \\
&\quad + \langle \mathscr{A}_N L, \mathscr{T}_X M\rangle + \langle \mathscr{A}_L M, \mathscr{T}_X N\rangle \\
\langle R(M,X)N,Y\rangle &= \langle (\nabla_M \mathscr{T})_X Y, N\rangle + \langle (\nabla_X \mathscr{A})_M N, Y\rangle \\
&\quad - \langle \mathscr{T}_X M, \mathscr{T}_Y N\rangle - \langle \mathscr{A}_M X, \mathscr{A}_N Y\rangle \\
\langle R(X,Y)Z,N\rangle &= \langle (\nabla_Y \mathscr{T})_X Z, N\rangle - \langle (\nabla_X \mathscr{T})_Y Z, N\rangle \\
\langle R(X,Y)Z,Z'\rangle &= \langle {}_pR(X,Y)Z,Z'\rangle - \langle \mathscr{T}_X Z, \mathscr{T}_Y Z'\rangle + \langle \mathscr{T}_Y Z, \mathscr{T}_X Z'\rangle \\
K(P_{MN}) &= {}_qK(P_{MN}) - 3\,\|\mathscr{A}_M N\|^2/\|M\wedge N\|^2 \\
K(P_{MX})\,\|M\|^2\,\|X\|^2 &= \langle (\nabla_M \mathscr{T})_X X, N\rangle + \|\mathscr{A}_M X\|^2 - \|\mathscr{T}_X M\|^2 \\
K(P_{XY}) &= {}_pK(P_{XY}) - \frac{\langle \mathscr{T}_X X, \mathscr{T}_Y Y\rangle - \|\mathscr{T}_X Y\|^2}{\|X\wedge Y\|^2}
\end{aligned}$$

For a path lying on a leaf, the holonomy of the canonical connection is equal to the foliation holonomy, so the foliation holonomy is orthogonal.

Proof. Since the canonical connection is basic ${}_qR(X,Y)={}_qR(X,N)=0$ by Proposition II.5.21. Choose an orthonormal frame composed of vector fields pulled back from the local base. Such a field is an infinitesimal automorphism, and so is the bracket of two such fields, which therefore differs from a pulled back vector field by a tangent vector field to the foliation. Hence,

$$\begin{aligned}
{}_qR(M,N)L &= \nabla_M \nabla_N L - \nabla_N \nabla_M L - \nabla_{[M,N]} L \\
&= \nabla_{\pi_* M} \nabla_{\pi_* N} \pi_* L - \nabla_{\pi_* N} \nabla_{\pi_* M} L - \nabla_{\pi_*[M,N]} L
\end{aligned}$$

as required. The formulas for the curvature operators and the sectional curvatures are proved by O'Neill [1966]. In the case of a Riemannian submersion, the tensors $\mathscr{T}$ and $\mathscr{A}$ introduced here differ from those of O'Neill only in the sign of $\mathscr{T}_X N$ and $\mathscr{A}_N X$. This changes only the formula for $\langle R(M,N)L,X\rangle$, and only the last three signs in this formula, but the calculations remain valid with appropriate sign changes throughout. Also, here the curvature of ${}_q\nabla$ is used instead of the curvature of the base, since this substitution is justified by the first conclusion of the proposition. Note that the last two formulas for R and the last formula for K are the classical equations of Codazzi and Gauss.

Since the canonical connection preserves the metric, its homogeneous holonomy group at any point is contained in the orthogonal group $O(q)$. For a path lying on a leaf, the foliation holonomy is the product of a finite number of germs belonging to the pseudogroup, that is, germs of isometries. Hence, for a closed path, it is an isometry with a fixed point, that is, an orthogonal transformation.

Since in general the holonomy of an adapted connection along a path lying on a leaf is the linear part of the foliation holonomy, and the foliation holonomy is linear, it must be equal to the connection holonomy. This completes the proof of Proposition 4.6. □

Before proceeding with the structure theory for Riemannian foliations, it is desirable to consider some of the important examples.

4.7 Example. The total space of any locally trivial smooth fibration admits a metric such that the foliation by components of fibers is a Riemannian foliation. Indeed, given any metric on the base space, any metric on the tangent bundle to the fibers, and any q-dimensional transversal plane field, there is a unique metric on the total space so that the given transversal field is normal and is mapped by the projection isometrically onto the base, while the metric induced on the fibers is the given one. The metric thus constructed satisfies condition (vi) of Proposition 4.2.

Many important examples of foliations arise as level sets of mappings or as orbits of group actions. The preceding example is the simplest case of Riemannian foliations arising from mappings, while the next few examples will show them arising from group actions. For the calculations, it is necessary to recall that a vector field X is an infinitesimal isometry (or Killing vector field) if and only if

$$\langle [X,Y],Z\rangle + \langle Y,[X,Z]\rangle = X\cdot\langle Y,Z\rangle \tag{4.8}$$

for all vector fields Y and Z.

4.9 Example. Let G be any Lie group with Lie algebra $\mathfrak{g}$ and $\mathfrak{h}$ any subalgebra of $\mathfrak{g}$. Choose a left invariant metric on G and orthonormal bases $\{X_a\}$ for $\mathfrak{h}$ and $\{N_\alpha\}$ for its orthogonal complement. Since $\mathfrak{h}$ is a subalgebra, it defines a foliation of G. Furthermore, the second fundamental form $\mathscr{S}$ of the normal plane field satisfies

$$\begin{aligned}\langle \mathscr{S}_{N_\alpha}N_\alpha, X_a\rangle &= \langle \nabla_{N_\alpha}N_\alpha, X_a\rangle = -\langle N_\alpha, \nabla_{N_\alpha}X_a\rangle\\ &= -\langle N_\alpha, \nabla_{X_a}N_\alpha\rangle - \langle N_\alpha,[N_\alpha,X_a]\rangle\\ &= \langle [N_\alpha,N_\alpha],X_a\rangle = 0.\end{aligned}$$

Since $\mathscr{S}_{N_\alpha}N_\beta$ is symmetric in (α,β), $\mathscr{S}=0$ and so the foliation is Riemannian. Since the leaf through the identity is a Lie subgroup H whose closure is also a Lie subgroup, the analysis of these examples reduces naturally to the two cases of dense subgroups and closed subgroups. An important special case is any foliation of a torus T^n by a subgroup of dimension p. The closure of each leaf is a torus T^m of dimension $m \geqq p$, and the tori T^m are the fibers of a fibration of T^n. The structure theory for arbitrary Riemannian foliations involves a generalization of this remark.

4.10 Example. Let G be a Lie group acting C^∞ as a group of isometries on the C^∞ Riemannian manifold M, and suppose all the orbits have the same dimension. Then the orbits are the leaves of a Riemannian foliation. To prove that they form a foliation, consider a point x in M and the isotropy subgroup G_x, consisting of all elements of G which leave x fixed. The Lie algebra $\mathfrak{g}$ of G, considered as a vector space, is the direct sum of the Lie algebra $\mathfrak{g}_x$ of G_x and a vector space $\mathscr{X}$. The action gives rise to a homomorphism from $\mathfrak{g}$ into the Lie algebra of vector fields on M, and near x, a basis for $\mathscr{X}$ maps into a basis for the tangent space to the orbits. Hence, the orbits form a C^∞ foliation. Since the second fundamental form $\mathscr{S}$ of the normal plane field is a tensor, to show that it vanishes at x, it suffices to show that any frame at x can be extended to a field of frames for which $\mathscr{S}$ vanishes. Given a normal frame $\{N_\alpha\}$ at x, extend is smoothly to points of a local cross-section, then to a neighborhood by radial flows in $\mathscr{X}$. It follows that at x, $\mathscr{L}_{X_a} N_\alpha = 0$. The argument from Example 4.9 then gives

$$\langle \mathscr{S}_{N_\alpha} N_\alpha, X_a \rangle = -\langle N_\alpha, [N_\alpha, X_a] \rangle = 0,$$

the last step of the argument in the preceding example being omitted because it depends on the additional fact that in that case, N_α is also a Killing vector field. In particular, whenever a compact Lie group acts smoothly, a Riemannian metric can be found for which it is a group of isometries. Indeed, given any metric, an invariant metric can be constructed by averaging with respect to Haar measure on the group. A simple example which is not a fibration is the action of S^1 on S^3 given in Example 2.45.

4.11 Example. Given a family $X(t)$, $t \in \mathbb{R}$, of nonsingular Killing vector fields on M, one gets a Riemannian foliation for a suitably chosen Riemannian metric on $M \times \mathbb{R}$. Consider the vector field $X(t) + \partial/\partial t$ and extend the metric $\langle\,,\rangle$ on M to $M \times \mathbb{R}$ by requiring

$$\langle X, \partial/\partial t \rangle = -\langle X, X(t) \rangle$$
$$\langle \partial/\partial t, \partial/\partial t \rangle = \langle X(t), X(t) \rangle$$

for all vectors X tangent to M. This is a bundle-like metric such that the quotient metric is the given metric on M. This example is due to Hermann [1979, Appendix M] who proposes it as a model for relativistic rigid motions.

4.12 Example. A principal bundle with group a Lie group G, being a smooth fibration, has the properties discussed in Example 4.7. It is interesting to inquire under what circumstances G acts as a group of isometries of the total space. The metric along the fibers can always be chosen so that G acts on each fiber by isometries. Indeed, in each local product neighborhood, one chooses a right invariant metric on the fiber, and then uses a partition of unity in the base to construct the metric globally. Since the left translation of a right invariant metric is a (possibly different) right invariant metric and a convex combination of right invariant metrics is right invariant, G preserves the metric thus constructed. Note that there is no relation between the metrics on two different fibers, unless a two sided invariant metric is used in the construction. If a transverse plane field invariant by the action of G can be found, then any metric on the base can be used to complete the construction of a metric on the total space invariant by G. If both a two-sided invariant metric on G and a transverse plane field invariant under G can be found, then in the metric constructed by using them, not only are the fibers isometric, but each one is totally geodesic. Indeed, let $\{X_a\}$ be the vector fields on the total space corresponding to a basis for the Lie algebra of G, and let $\{N_\alpha\}$ be a local framing of the normal bundle to the fibers. Then

$$\begin{aligned}\langle \nabla_{X_a} X_a, N_\alpha \rangle &= -\langle X_a, \nabla_{X_a} N_\alpha \rangle \\ &= -\langle X_a, \nabla_{N_\alpha} X_a \rangle - \langle X_a, [X_a, N_\alpha] \rangle = 0,\end{aligned}$$

so that by symmetry the second fundamental form of the fibers vanishes. In particular, if G is compact both the two-sided invariant metric and the invariant plane field can be constructed by averaging.

4.13 Example. If a mapping $f\colon M \to B$ of Riemannian manifolds is a Riemannian submersion, then M is foliated by the components of the level sets of f, and the foliation is Riemannian by condition (vi) of Proposition 4.2. If M is complete, then B is also, and f is a locally trivial fiber bundle over B. As observed by Hermann [1960], both these remarks follow from the relation between geodesics on B and orthogonal geodesics on M. He also proves that if the leaves are totally geodesic, the group of this bundle can be reduced to a Lie group, namely, the group of isometries of the fiber. The idea of the proof is to examine the map from one fiber to another obtained by using the normal lifts of a path in B. This map is an isometry because of the relation between the second fundamental form of a submanifold and the first normal variation of arc length of a curve lying in the submanifold. This example and the previous one can be combined to yield the result that a locally trivial smooth fiber space with compact fiber defines a Riemannian foliation in which the fibers are totally geodesic if and only if its group can be reduced to a compact Lie group.

These examples show that there are enough Riemannian foliations to make the subject interesting. On the other hand, there is a rather explicit structure theorem. As is well-known, the bundle of orthonormal frames on a Riemannian manifold admits a canonical parallelization, constructed by using the Levi-Cività connection and the tautological form. The analogous construction on the normal frame bundle of a Riemannian foliation, using the canonical connection and the tautological form, parallelizes the normal bundle to the lifted foliation. The crucial fact is that each of the vector fields in this parallelization is an infinitesimal automorphism. Since any vector field tangent to the lifted foliation is also an infinitesimal automorphism, on a compact manifold the group of automorphisms is transitive. This high degree of symmetry of the lifted foliation is what makes the structure theorem possible. To state it precisely, some further definitions are needed.

4.14 Definition. A vector field on a foliated manifold is *projectable* if its normal component is locally the pullback of a vector field on the local base. Given a Lie algebra $\mathfrak{g}$, a *Lie* $\mathfrak{g}$*-foliation* is a foliation together with a trivialization of the normal bundle by projectable vector fields which make up a Lie algebra isomorphic to $\mathfrak{g}$.

4.15 Proposition. *A vector field on a foliated manifold is projectable if and only if it is an infinitesimal automorphism of the foliation.*

Proof. A vector field is projectable if and only if in local coordinates (x^a, y^α) it has the form

$$f^a(x, y)\,\partial/\partial x^a + g^\alpha(y)\,\partial/\partial y^\alpha.$$

However, this is precisely the condition that its Lie bracket with every tangent vector be a tangent vector. □

In order to state the next definition, let ${}_q\theta^\alpha$ denote the tautological form on the bundle of orthonormal frames of the normal bundle of a foliation, and let ${}_q\omega^\alpha_\beta$ denote the connection form of a connection in this bundle. Then ${}_q\theta^\alpha + {}_q\omega^\alpha_\beta$ is a form with values in the vector space $\mathbb{R}^q \oplus \mathfrak{o}(q)$. Indeed, given a basis $\{e_\alpha\}$ in $\mathbb{R}^q$, let e^β_α denote the $q \times q$ matrix with $+1$ in the (α, β) entry, -1 in the (β, α) entry, and 0 otherwise. Then any vector A tangent to the frame bundle is mapped into

$${}_q\theta^\alpha(A)\,e_\alpha + {}_q\omega^\alpha_\beta(A)\,e^\beta_\alpha$$

and this map is an isomorphism from the normal space of the lifted foliation to $\mathbb{R}^q \oplus \mathfrak{o}(q)$. An inverse is well-defined by requiring that e_α map to a horizontal vector and that e^β_α map to the corresponding fundamental vector field for the frame bundle. This gives a parallelization of the normal frame bundle to the lifted foliation. It will be convenient to represent each element of the parallelization by a vector field on the normal frame bundle, defined up to tangent fields to the lifted foliation.

4.16 Definition. Given a Riemannian foliation, the *canonical parallelization* of the normal bundle to the lifted foliation is the parallelization associated to the canonical connection by the preceding construction.

4.17 Lemma. *Given a Riemannian foliation, the vector fields belonging to the canonical parallelization of the normal bundle to the lifted foliation are infinitesimal automorphisms of the lifted foliation.*

Proof. Any local projection of a Riemannian foliation induces a local projection of its normal frame bundle onto the tangent frame bundle of the model space, and by Proposition II.5.14, the lifted foliation is the inverse image of the point foliation of the tangent frame bundle. Since the canonical connection and ${}_q\theta$ are also pullbacks, the vector fields belonging to the canonical parallelization are projectable. The result follows by Proposition 4.15. □

4.18 Theorem. *Given a Riemannian foliation of a compact manifold, its normal frame bundle V is fibered by the closures of the leaves of the lifted foliation. This fibration $\pi\colon V \to W$ is locally trivial, and the restiction of the lifted foliation to each fiber is a Lie $\mathfrak{g}$-foliation, where $\mathfrak{g}$ depends neither on the fiber nor on the Riemannian structure. Moreover, there is a locally trivial sheaf over V whose fiber consists of germs of infinitesimal automorphisms which commute with all the global infinitesimal automorphisms of the lifted foliation. This fiber is a Lie algebra anti-isomorphic to $\mathfrak{g}$.*

Proof. See Molino [1977]. A sketch of his proof follows. Since the given manifold and the orthogonal group $O(q)$ are both compact, V is compact. Thus, Lemma 4.17 implies that the automorphism group of the foliation is transitive. It follows that the number of independent functions constant on the leaves (basic functions) is the same at any point, so the level sets of these functions form a locally trivial fibration whose fibers are the closures of leaves of the lifted foliation. Each such fiber is itself foliated by a subset of these leaves, in such a way that its normal bundle in the fiber is parallelized by infinitesimal automorphisms $N_1, \ldots, N_r$. $[N_i, N_j]$ is a linear combination of the N_k with basic, hence constant, coefficients, so $\{N_1, \ldots, N_r\}$ spans a Lie algebra and the foliation on each fiber of π is a Lie foliation. By the transitivity of the automorphism group, the algebras for different fibers are isomorphic, and the sheaf defined by the commutation property is locally trivial. On a Lie group, the set of vector fields which commute with all left invariant fields is precisely the set of right invariant vector fields, which form a Lie algebra anti-isomorphic to the algebra of left invariant fields. This fact generalizes immediately to the case at hand, so that the sheaf in question can be viewed as a generalization of the right action of a group on a principal bundle. ▯

In view of this theorem, it is useful to recall the structure theorem for Lie foliations.

4.19 Proposition. *Let M be a compact manifold admitting a Lie $\mathfrak{g}$-foliation $\mathscr{F}$, and let G be the connected, simply connected Lie group corresponding to $\mathfrak{g}$. Then there is a regular covering $\tilde{M}$ of M such that the leaves of the induced foliation $\tilde{\mathscr{F}}$ of $\tilde{M}$ are fibers of a locally trivial fibration of $\tilde{M}$ over G. Furthermore, the closures of the leaves of $\mathscr{F}$ are fibers of a locally trivial fibration of M.*

Proof. See Fedida [1971]. ▯

Because of the high degree of symmetry, the lifted foliation consists of isomorphic leaves with trivial holonomy. Furthermore, each lifted leaf is a covering space of some leaf of the original foliation. Though the lifted leaf has higher codimension, its holonomy along any closed path determines the holonomy along the projected path. Since the holonomy along the projected path is orthogonal, it also determines the holonomy along the lifted path. Thus, the lifted leaf is the holonomy covering of the given leaf, which proves the following result.

4.20 Proposition. *For a Riemannian foliation $\mathscr{F}$ of a compact manifold, any leaf of the lifted foliation is diffeomorphic to the holonomy covering of each leaf of $\mathscr{F}$.*

4.21 Proposition. *For a Riemannian foliation of codimension q with compact leaves, each leaf has finite holonomy and is covered by all nearby leaves. The quotient space is locally homeomorphic to the quotient of $\mathbb{R}^q$ by a finite subgroup.*

Proof. Since the leaves are compact, the holonomy group is a subgroup of O_q with the property that each orbit under its natural action on the normal sphere S^{q-1} is finite. It follows that the identity component of the closure in O_q of the holonomy group consists of a single element, so the holonomy group is finite. The mapping which takes S^{q-1} to its center induces the covering of nearby leaves to a given leaf. The last statement follows immediately. □

Some remarks can also be made about the quotient space for arbitrary Riemannian foliations of a compact manifold. Let $\pi: V \to W$ be the fibration of the normal frame bundle with fibers the closures of leaves of the lifted foliation. Then the natural action of O_q on V induces an action of O_q on W such that the quotient is the quotient space of the original foliation by the closures of its leaves. Combining this idea with the formulas for the bracket of the canonical parallelization leads to structure theorems in many special cases (Molino [1977]). The case of Riemannian flows (1-dimensional foliations) has been studied by Carrière [1981], while the case of parallel curvature has been studied by Blumenthal [1983].

The graph of a Riemannian foliation has particularly nice properties which make it useful for geometric results. These will be stated after a preliminary definition is given.

4.22 Definition. A Riemannian submersion is *horizontally complete* if each horizontal geodesic can be extended as far as desired in both directions.

4.23 Proposition. *Let $\mathscr{F}$ be a Riemannian foliation of the manifold M. Then its graph Γ is a Hausdorff space. Γ admits a unique Riemannian metric such that A and Z are Riemannian submersions. With respect to this metric, T is an isometry, each of the three foliations of Γ is Riemannian, and each leaf of the $2p$-dimenional foliation is locally the Riemannian product of leaves of the p-dimensional foliations. If M is compact or A is horizontally complete, then Γ is complete. If A is horizontally complete, then it is a locally trivial fibration, and its fiber is the holonomy covering of each leaf of $\mathscr{F}$. If M is compact, this fiber is diffeomorphic to a leaf of the lift of $\mathscr{F}$ to its normal frame bundle. If M and all the leaves are compact, then Γ is compact.*

Proof. Since the holonomy is orthogonal, it is linear and therefore analytic. Thus, by Proposition 2.45, Γ is a Hausdorff space. Since $Z \oplus A$ is an immersion of Γ into $M \times M$, the pullback of the product metric on $M \times M$ is a metric on Γ. With respect to this metric, A and Z are submersions, and clearly there can be at most one metric with this property. Since interchanging factors in $M \times M$ is an isometry, T is an isometry. Since A and Z are submersions, each p-dimensional foliation is Riemannian. Since the given foliation $\mathscr{F}$ is Riemannian, its pullback by A, which is the $2p$-dimensional foliation, is Riemannian. Locally, each leaf of the $2p$-dimensional foliation inherits the product metric from $M \times M$. By Proposition 3.6, if M is complete, then each leaf is complete, so each covering space of a leaf, thus each fiber of A is complete. If M is compact, then there is an $\varepsilon > 0$ such that each point of M belongs to a geodesic ball of radius ε. Thus, each point of $M \times M$ and each point of Γ belongs to a geodesically convex set of diameter at least 2ε. Let $\{x_n\}$ be a Cauchy sequence in Γ. Then $A(x_n)$ is Cauchy, so converges to a point y in M. For n large enough, there is a unique shortest geodesic joining $A(x_n)$ to y, and there is a unique horizontal lifting of this geodesic to a geodesic joining x_n to a point y_n in $A^{-1}(y)$. The sequence $\{y_n\}$ is also Cauchy,

$$d(y_n, y_m) \leqq d(y_n, x_n) + d(x_n, x_m) + d(x_m, y_m)$$

and $d(y_n, x_n) = d(y, A(x_n))$. (Here $d(,)$ denotes the topological metric associated to a Riemannian metric.) Hence $\{y_n\}$ converges to a point x which is also the limit of $\{x_n\}$. If A is horizontally complete, then M is complete and a slight modification of the preceding proof shows that Γ is complete. If A is horizontally complete, then a product structure can be constructed over the neighborhood of any point of M by using horizontal lifts of a family of radial geodesics. By Proposition 2.44, the fiber of A is diffeomorphic to the holonomy covering of each leaf of $\mathscr{F}$, so by Proposition 4.20 it is also diffeomorphic to a leaf of the lifted foliation. If M is compact, then Γ is complete, so A is locally trivial. If all the leaves are compact, then by Proposition 4.21 each leaf has finite holonomy, so the fiber of A is compact and therefore Γ is compact. This concludes the proof of the proposition. □

4.24 Definition. If A is horizontally complete, its fiber is called the *universal leaf* of $\mathscr{F}$.

4.25 Corollary. *For a Riemannian foliation $\mathscr{F}$ of a compact manifold with all leaves compact, both the order of the holonomy groups and the volume of the leaves are bounded.*

Proof. Since the holonomy covering of a leaf factors through the holonomy covering of nearby leaves, the order of the holonomy group is locally bounded, hence bounded because M is compact. Though the fibers of A are not isometric, their volume is bounded, and this bound also bounds the volume of the leaves of $\mathscr{F}$. □

Another application of the graph is based upon the curvature relations for a Riemannian submersion.

4.26 Theorem. *Suppose a Riemannian manifold with nonpositive sectional curvature admits a Riemannian foliation such that the linear holonomy maps h satisfy*

$$\langle h(Q\nabla_X X), Q\nabla_Y Y\rangle \geqq 0$$

for every pair of tangent vector fields X and Y. Then the graph has nonpositive sectional curvature. If the manifold is compact, then the universal cover of the universal leaf is contractible.

Proof. See the paper of Winkelnkemper [1982]. □

Since the curvature of a Riemannian foliation has stronger vanishing properties than that of an arbitrary foliation, its characteristic classes should have special properties. Indeed, many of the usual classes vanish, but there are additional classes which can be defined only for Riemannian foliations.

4.27 Proposition. *The WO_q characteristic classes of a Riemannian foliation vanish, except for the Pontrjagin classes of the normal bundle. Those W_q characteristic classes which are divisible by C_j for some odd j vanish for a Riemannian foliation.*

Proof. According to Definition II.5.23, the WO_q classes are represented by products of forms h_i and C_j, where the C_j represent characteristic classes of the normal bundle and the h_i are defined by comparing a Riemannian connection with an adapted connection. For a Riemannian foliation, the canonical connection is both Riemannian and adapted, and the h_i defined by comparing it with itself are 0. The W_q classes are obtained by comparing an adapted connection with the connection associated to a trivialization. If the adapted connection is also Riemannian, then the corresponding C_j for j odd vanish. This completes the proof. □

4.28 Definition. Let $R\Gamma_q$ denote the groupoid of germs of isometries of Riemannian manifolds, and let $SR\Gamma_q$ be the subgroupoid consisting of oriented germs. Let $FR\Gamma_q$ denote the fiber of the fibration

$$v\colon BR\Gamma_q \to BO_q$$

where v is induced by the map which takes a germ to its Jacobian matrix. Then the *relative characteristic classes* of a Riemannian foliation are those associated with the cohomology of $BR\Gamma_q$, while the *absolute characteristic classes* of a Riemannian foliation with trivial normal bundle are those associated with the cohomology of $FR\Gamma_q$.

All of these definitions are in the general context of Γ-structures, as discussed in § III.1. Note that $FR\Gamma_q$ is also the fiber of the fibration

$$v\colon BSR\Gamma_q \to BSO_q$$

because a foliation with trivial normal bundle has a natural transverse orientation. The method of comparison of connections can be extended to define certain absolute classes, by using the canonical connection and the connection associated to an orthonormal framing. (This method does not apply to the relative classes because there is only one natural connection available.) For this purpose, consider the cochain complex RW_q defined for q even by taking as the ring of cochains

$$\text{(4.29a)} \qquad \mathbb{R}[C_\chi, C_2, \ldots, C_{q-2}]_q \otimes \Lambda(h_\chi, h_2, \ldots, h_{q-2})$$

with the differential

$$\text{(4.29b)} \qquad dC_j = 0 \qquad dh_j = \begin{cases} C_j & j = \chi, 2, 4, \ldots, \frac{1}{2}q \\ 0 & j > \frac{1}{2}q \end{cases}$$

C_χ has dimension q and h_χ has dimension $q-1$. For q odd, C_χ and h_χ are omitted and the largest index is $q-1$. C_χ and h_χ are introduced because SO_q for q even has an invariant polynomial, corresponding to the Euler class, that does not exist for GL_q. (As a cochain on the Lie algebra, C_χ corresponds to the square root of the determinant.) Since both connections are Riemannian, C_j and h_j vanish for j

odd. Finally, the truncation index q corresponds to the stronger vanishing theorem for the curvature (Proposition 4.6).

4.30 Proposition. *Given any manifold M with a Riemannian foliation with trivial normal bundle, there is a cochain mapping from RW_q into the de Rham complex of M, hence a homomorphism from $H^*(RW_q)$ to $H^*(M; \mathbb{R})$. This homomorphism is the composition of a fixed homomorphism from $H^*(RW_q)$ to $H^*(FR\Gamma_q)$ and the homomorphism from $H^*(FR\Gamma_q)$ to $H^*(M)$ induced by the classifying map. The classes in dimension greater than q do not depend upon the choice of metric. If $\phi\colon M \to SO_q$ is a change of framing, then there is a formula for the change in characteristic classes that depends on $\{\phi^* \tau C_i\}$ where τ is the transgression homomorphism in the universal SO_q-bundle. $H^*(RW_q)$ is generated by elements of the following forms:*

$$h_j \qquad \tfrac{1}{2}q < j < q$$

$$C_{i_1}\dots C_{i_r} \quad h_{j_1}\dots h_{j_s} \begin{cases} i_1 \leqq \dots \leqq i_r \\ j_1 < \dots < j_s \\ I = \dim(C_{i_1}\dots C_{i_r}) \leqq q \\ I + \dim h_{j_1} > q. \end{cases}$$

The homomorphism

$$v^*\colon H^r(BSO_q) \to H^r(BSR\Gamma_q)$$

is trivial for $r > q$ and real coefficients, but is injective for $q \geqq 2$ and coefficients, $\mathbb{Z}/p\mathbb{Z}$, where p is an odd prime. $\pi_1(BSR\Gamma_1)$ is isomorphic to $\mathbb{R}$, while its higher homotopy groups vanish. $FR\Gamma_q$ is $(q-1)$ connected, while $\pi_q(FR\Gamma_q)$ admits an epimorphism onto $\mathbb{R}$. The classes in dimension greater than $q+1$ remain constant under continuous variations. The homomorphism from $H^*(WR_q)$ to $H^*(FR\Gamma_q)$ is injective, and if q is congruent to 2 or 3 modulo 4, all the variable classes can vary independently.

Proof. Most of these results are contained in the papers of Pasternack [1973], Lazarov and Pasternack [1976], and Hurder [1981 a, c]. The proofs are generally analogous to those of the corresponding statements for Γ_q. The calculation of the homotopy of $BSR\Gamma_1$ is based upon the observation that an $SR\Gamma_1$ structure is equivalent to a closed 1-form. The calculation of $\pi_q(FR\Gamma_q)$ is also related to the existence of closed transverse q-forms, or transverse volume forms. □

One of the powerful tools in Riemannian geometry is the theory of elliptic partial differential operators. There are many such operators defined on any Riemannian manifold, but in the case of a Riemannian foliation there are also transversally elliptic operators and tangentially elliptic operators to be considered. One important class of elliptic operators is obtained by generalizing the theory of harmonic maps of Riemannian manifolds. Examples of such maps are harmonic functions, the injection map of a minimal submanifold, and a Riemannian submersion such that the inverse image of each point is a minimal submanifold. It is this last example that offers an interesting generalization to foliations. As usual, the challenge is to find the correct method of overcoming the lack of a reasonable global quotient space for a foliation. In this case, one must consider forms on the total space with values in the normal bundle, with particular attention to the projection operator Q, which may be regarded as a 1-form with values in the normal bundle.

Given a vector bundle E with a connection ∇, the exterior derivative d_∇ for E-valued r-forms is defined by

$$(d_\nabla \omega)(X_1, \ldots, X_{r+1}) = \sum_{i=1}^{r+1} \nabla_{X_i} \omega(X_1, \ldots, \hat{X}_i, \ldots, X_{r+1})$$
$$+ \sum_{i<j} (-1)^{i+j} \omega([X_i, X_j], X_1, \ldots, \hat{X}_i, \ldots, \hat{X}_j, \ldots, X_{r+1})$$

and satisfies

$$d_\nabla^2 = R_\nabla \wedge \omega$$

where R_∇ is the curvature of ∇. If E is a bundle over a Riemannian-manifold M^n, then the operator * can be defined for E-valued forms, and the formal adjoint to d_∇ is defined for an r-form by

$$d_\nabla^* = (-1)^{n(r+1)+1} * d_\nabla *.$$

If E is the normal bundle to a foliation, ∇ is the adapted connection consistent with the metric of M, and Q is the projection operator of the tangent bundle of M onto the normal bundle of the foliation, then $d_\nabla Q = 0$ and $d_\nabla^* Q$ is the mean curvature vector of the leaves. Two second order operators of interest are the Laplacian Δ, defined for Q-valued forms by

$$\Delta = d_\nabla d_\nabla^* + d_\nabla^* d_\nabla$$

and the Jacobi operator J_∇, defined for normal vector fields L by

$$J_\nabla(L) = d_\nabla^* d_\nabla \mathrm{L} + \sum_{\alpha=1}^{q} \langle R_\nabla(N_\alpha, L) N_\alpha, L \rangle$$

where $\{N_\alpha\}$ is an orthonormal basis for the normal bundle. L is called a Jacoby field if $J_\nabla(L) = 0$, and the r-form ω is called harmonic if $d_\nabla \omega = 0$ and $d_\nabla^* \omega = 0$. The energy functional is defined on a compact oriented manifold M by

$$\mathscr{E}(\omega) = \tfrac{1}{2} \int_M g_Q(\omega \wedge *\omega)$$

where $\omega \wedge *\omega$ is regarded as having as values pairs of normal vectors, thus is in the domain of the metric on the normal bundle.

4.31 Proposition. *Given a foliation of a Riemannian manifold M, the leaves are minimal if and only if Q is harmonic. If the metric is bundle-like, and the manifold is compact and oriented, then a form ω is harmonic if and only if $\Delta\omega = 0$. Furthermore, Q is harmonic if and only if it is an extremal of the energy functional $\mathscr{E}$. If the Ricci curvature of the local base is nonpositive everywhere and negative somewhere, then the given foliation is a local minimum for the energy functional under a certain class of normal variations, and every infinitesimal metric automorphism of the foliation is tangential. If a Riemannian foliation admits a metric such that the leaves are minimal, then it admits a bundle-like metric for which the leaves are minimal.*

Proof. See Kamber and Tondeur [1982]. The Jacobi fields defined above occur in the study of the second variation of the energy functional. If the metric is not bundle-like, there are additional terms in the trace formula defining J_∇ that prevent the theory from being carried through. □

There are also some interesting differential operators transverse to the foliation, which are in some sense elliptic in the transverse direction only. In order to define these, it is convenient to look first at some properties of differential forms on a foliated Riemannian manifold. Because the metric defines a plane field complementary to the foliation, namely the orthogonal plane field, the foliation has a natural underlying $GL_{p,q;m}$-structure. Thus, Definition II.5.11 applies, and each k-form is uniquely decomposed as the sum of forms of type (r, s), where $r + s = k$. Of particular interest are the forms ϕ of type $(0, s)$, such that $d\phi$ is the type $(0, s + 1)$. In local coordinates, such a form can be written

$$\phi_{j_1 \ldots j_s}(y)\, dy^{j_1} \wedge \ldots \wedge dy^{j_s}$$

so that locally it is the pullback of a form on the local base. Note that the exterior derivative of such a form satisfies the same conditions, so there is a cohomology theory defined.

4.32 Definition. On a foliated Riemannian manifold $(M, \mathscr{F})$, a differential form ϕ of type $(0, s)$ such that $d\phi$ is of type $(0, s + 1)$ is called a *basic s-form.* The restriction of the exterior derivative to the basic forms is denoted by d''. The cohomology of the basic forms under d'' is called the *basic cohomology* and denoted by $H_b^*(M, \mathscr{F})$.

The property of being basic depends only on $\mathscr{F}$ and not on the choice of metric, so the basic cohomology depends only on $\mathscr{F}$. G. Schwartz [1974] has shown that on a compact manifold, H_b^0 and H_b^1 are finite dimensional, but in higher dimensions H_b^* can be infinite dimensional.

If the metric is bundle-like, then the operator $*''$ defined locally by using the metric on the local base is in fact globally defined and maps basic forms to basic forms. Thus d'' has a formal adjoint defined on basic s-forms by

$$\delta'' = (-1)^{qs+q+1} *'' d'' *''.$$

The basic Laplacian is the operator Δ'' defined by

$$\Delta'' = d''\delta'' + \delta'' d''.$$

A basic form ϕ is harmonic if it satisfies $d''\phi = 0$ and $\delta''\phi = 0$. One expects that d'' and δ'' should be extendable to densely defined, closed operators on some Hilbert space with domains such that the extended operators are adjoint. At present, no such extension is known. To see some of the difficulties, it suffices to consider the basic q-form obtained locally by pulling back the Riemannian volume form on the local base. This is certainly a closed form, but the following example, due to Carrière, shows that it can be trivial in basic cohomology.

4.33 Example. Consider the compact M^3 obtained as the quotient of $T^2 \times \mathbb{R}$ by the equivalence relation $(x, t) \sim (A(x), t + 1)$, where A is an element of $SL_2(\mathbb{Z})$

with trace greater than 2. Consider the vector field tangent to each $T^2 \times \{t\}$ obtained by translating an eigenvector of A with eigenvalue less than 1. This vector field is not a Killing vector field, but it does define a Riemannian foliation of codimension 2 such that $H_b^2 = 0$. In particular, the basic Riemannian volume form is exact. In this example, the basic cohomology is isomorphic to the cohomology pulled back from S^1 by the projection of M^3 onto S^1 induced by the projection of $T^2 \times \mathbb{R}$ onto $\mathbb{R}$.

One can now return to the question of adjointness for d'' and δ''. Since the volume form of a Riemannian manifold is harmonic, the basic volume form σ satisfies $d''\sigma = 0$ and $\delta''\sigma = 0$. Since also $\sigma = d''\psi$, adjointness in some Hilbert space would imply

$$\langle \sigma, \sigma \rangle = \langle d''\psi, \sigma \rangle = \langle \psi, \delta''\sigma \rangle = 0,$$

a contradiction since $\sigma \neq 0$. Note that for real cohomology and harmonic forms on a compact Riemannian manifold with non-empty boundary, the same problem arises. In that case, the adjoint to d'' is obtained by first restricting δ'' to forms whose support is contained in the interior of the manifold, then extending this restriction to a closed operator. Thus, σ is omitted from the domain and no contradiction arises.

There are a number of types of examples where the basic volume form is known to be nontrivial in basic cohomology. In each of these examples, the manifold is compact and the basic cohomology satisfies a Poincaré duality theorem with respect to the dimension q. The first such example is the local product structure studied by Reinhart [1958]. Another example is a complex analytic foliation with compact leaves and a Hermitian bundle-like metric (Girbau and Nicolau [1980]). The third example involves certain transversely homogeneous foliations, that is, foliations defined by submersions onto a homogeneous space G/K with coordinate transformations belonging to G. In this case, Blumenthal [1980] has shown that the basic cohomology is isomorphic to the cohomology of G/K. Note that in some of these examples, the global quotient space may be a very bad topological space. If the total space is complete but not compact, the natural object of study is differential forms which are square integrable (Kitahara [1979]).

Instead of transverse operators, one may consider operators elliptic along the leaves. Theorem 2.51 and its generalizations give index theorems for operators elliptic along the leaves. For these theorems, one needs a transverse invariant measure, and the formula in fact depends upon the choice of measure. Since a Riemannian foliation admits a canonical transverse invariant measure, namely, the basic volume form, there is a well-defined index theorem for a given operator elliptic along the leaves. In those cases, such as Example 4.33, where the basic volume form is exact, the indices of the operators must all vanish, since the foliation cycle on which the characteristic classes are evaluated vanishes. In particular, the component along the leaves of the exterior derivative defines a complex of differential operators that is elliptic along the leaves. This component along the leaves is well-defined, by the following proposition.

4.34 Proposition. *Given a foliation with a fixed complementary plane field, the exterior derivative of a form of type* (r, s) *is the sum of terms of types* $(r + 1, s)$,

$(r, s+1)$, and $(r-1, s+2)$. The operator

$$d' = \underset{1,0}{\pi}\, d$$

that takes each form of type (r, s) to a form of type $(r+1, s)$ is independent of the choice of complement. Its expression in local coordinates is given by

$$(d'\phi)_{k_1 \ldots k_{r+1} j_1 \ldots j_s} = \delta^{i i_1 \ldots i_r}_{k_1 \ldots k_{r+1}} \partial\phi_{i_1 \ldots i_r} / \partial x^i,$$

modulo terms of degree $(s+1)$ or more in dy^α. d' defines a complex of differential operators elliptic along the leaves.

Proof. Examination of the formula for $d\phi(A_1, \ldots, A_{r+s+1})$ in terms of values of ϕ shows that if ϕ is of type (r, s) and the first plane field is integrable, the only nonzero components have types $(r+1, s)$, $(r, s+1)$, and $(r-1, s+2)$ as required. In a local coordinate system, choose a basis

$$\{\theta^a = dx^a + A^a_\alpha(x, y)\, dy^\alpha, dy^\alpha\}$$

for the 1-forms which is consistent with some given splitting. Then any form of type (r, s) can be written, modulo terms with $(s+1)$ or more dy^α,

$$\phi = \phi_{i_1 \ldots i_r j_1 \ldots j_s} dx^{i_1} \wedge \ldots \wedge dx^{i_r} \wedge dy^{j_1} \wedge \ldots \wedge dy^{j_s}.$$

Moreover, under any change of coordinates, these initial terms determine each other. The local formula for d' follows immediately, by calculating d in a coordinate system. Furthermore, the local formula is independent of the choice of complement. Since the only component of $d^2\phi$ of type $(r+2, s)$ is given by $(d')^2\phi$, it follows that $(d')^2 = 0$. The exactness along the leaves of the symbol sequence for d' follows from the coordinate formula and the usual calculation for d. This completes the proof of the proposition. □

The harmonic spaces mentioned in the statement of Theorem 2.51 arise from the d'-complex by use of a Riemannian metric on the tangent bundle to the leaves. As with the Euler number of a manifold, the operator whose index must be computed is $d' + (d')^*$, where $(d')^*$ is the adjoint of d', because the Laplacian

$$\Delta' = d'(d')^* + (d')^* d'$$

is self-adjoint and therefore has index 0.

Let $\mathscr{B}^s$ denote the sheaf of germs of basic s-forms, and note that such forms are characterized by the property of being d'-closed forms of type $(0, s)$. Then Vaisman [1971] has shown that $H^r(M; \mathscr{B}^s)$ is isomorphic to the d'-cohomology of forms of type (r, s). He also considers the nonelliptic second order operator Δ_f obtained from d' and its adjoint with respect to a full metric on a compact manifold. For forms of (r, s) type, he shows that the space of forms harmonic in the usual sense is contained in the space of Δ_f-harmonic forms, which is contained in the d'-cohomology space. Unfortunately, the (r, s) component of a usual harmonic form is not in general harmonic, and both containments may be proper.

For the quotient bundle of a foliation, the transition functions between coordinate systems are basic with respect to the induced foliation of the coordinate system. Thus it is natural to consider the class of vector bundles with this property which are called foliated vector bundles. The obstructions to the existence of basic connections for such bundles, and more generally to the existence of basic differential operators with a given symbol, have been studied by Molino [1971], Kamber and Tondeur [1971], and Vaisman [1974].

Conformal structures are so closely related to Riemannian structures that it is natural to end this section with a discussion of conformal foliations. Let $C(q)$ denote the conformal group of euclidean q-dimensional space, that is, the group of linear transformations that preserve angles. $C(1)$ is the same as $GL(1,\mathbb{R})$ and the orientation preserving subgroup of $C(2)$ can be identified with $GL(1,\mathbb{C})$, but for $q \geqq 3$ this group is not much larger than $O(q)$. An element of $C(q)$ is characterized among linear transformations by the fact that the metric is multiplied by a positive constant, rather than being preserved. For nonlinear conformal transformations, such a constant is defined at any point, but for $q \geqq 3$, it is independent of the point. The study of conformal foliation begins with the following proposition.

4.35 Proposition. *For a foliated manifold, the following statements are equivalent:*

(i) *There is a Riemannian metric which is locally conformal to a bundle-like metric.*

(ii) *The foliation admits a conformal Γ-structure.*

(iii) *There is a reduction of the group of the normal bundle to $C(q)$, such that the transition functions in the reduced bundle are basic with respect to the induced foliation of the coordinate system.*

(iv) *There is an adapted connection on the normal bundle whose holonomy group at some point is contained in $C(q)$.*

(v) *There is a metric g such that with respect to the adapted connection $\nabla^{\mathcal{T}}$*

$$(\nabla_X^{\mathcal{T}} g)(M,N) = \lambda(X)\, g(M,N)$$

where λ is the 1-form defined by

$$\lambda = \frac{1}{q} d' \ln \det(h_{\alpha\beta})$$

and $h_{\alpha\beta}$ is the normal part of g.

Furthermore, g is globally conformal to a bundle-like metric if and only if λ is d'-exact.

Proof. See Vaisman [1979]. □

Note that any foliation of codimension 1 is conformal, as is any complex analytic foliation of complex codimension 1. With regard to the characteristic classes of a conformal foliation, properties similar to those of a Riemannian foliation may be expected.

4.36 Proposition. *For a conformal foliation in codimension at least 3, the normal bundle admits a basic connection, so the Chern forms vanish in dimension greater than q. For codimension 1 or 2, the Pontrjagin classes vanish above dimension q, even though the Chern forms may not.*

Proof. The first statement is proved by Nishikawa and Sato [1976], who also discuss projective foliations of codimension at least 2. The second statement is proved by Montesinos [1982]. □

5. Foliations with a Few Derivatives

In this section are collected a number of important results about C^1 and C^2 foliations, mostly in codimension 1. The proofs are analytic, but in some sense elementary. Indeed, the theorems are obtained by sophisticated and nontrivial use of the mean value theorem. For C^1 foliations, the result in question is Thurston's [1974b] generalization of the Reeb stability theorem to compact leaves with nonvanishing first Betti number. For C^2 foliations in codimension 1, the results considered are Kopell's [1970] theorem about commuting germs and two theorems of Sacksteder [1965], a fixed point theorem for the holonomy pseudogroup and an existence theorem for transverse invariant measures for foliations without holonomy. The fixed point theorem was originally introduced as a lemma to the existence theorem, for which there now exists a much more intuitive proof not using this lemma. Meanwhile, the fixed point theorem has become very important in the structure theory of codimension 1 foliations. Indeed, it implies that on a compact manifold, the complement of a minimal set contains a set which is minimal for the induced foliation. Since this complement is noncompact, there is no analogous result in the usual purely topological theory of minimal sets for group actions.

In addition to the theorems obtained by analytic arguments, there are important geometric results of Novikov [1965], stated by him for C^∞ foliations, but valid in the C^2 case by the same proofs. The section ends with these results and various applications of the main theorems.

5.1 Theorem. *Let $\mathscr{F}$ be a C^1 codimension q foliation, L a compact leaf of $\mathscr{F}$, and H the holonomy homomorphism. Then at least one of the following statements is true:*

(i) *$dH\colon \pi_1(L) \to GL_q$ is nontrivial.*
(ii) *$H^1(L; \mathbb{R})$ is nontrivial.*
(iii) *H is trivial and L has a saturated tubular neighborhood with a product foliation. If $q = 1$, the manifold is compact, and $\mathscr{F}$ is transversally oriented, then $\mathscr{F}$ is a fibration over S^1 or I.*

This theorem is proved by Thurston [1974b], who also gives an example to show that the result is false for C^0 foliations unless the fundamental group is finite, in which case the original Reeb [1952] stability theorem applies. The proof of the theorem consists of Reeb's arguments and the following lemma.

5.2 Lemma. *Let G be a nontrivial, finitely generated group of germs at 0 of C^1 diffeomorphisms of $\mathbb{R}^q$ such that each element of G is tangent to the identity. Then G admits a nontrivial homomorphism into $\mathbb{R}^q$.*

Proof. Very short proofs have been given by Schachermayer [1978], and by Reeb and Schweitzer [1978]. Thurston motivates the proof by remarking that a germ with derivative vanishing on a neighborhood is a translation, hence if the derivatives vanish at a point it should be a translation in the limit. All the proofs are in some sense ad hoc, and the homomorphisms constructed have no naturality properties. On the other hand, if there exists a least finite k such that some element of G has a nontrivial k-jet, then G admits a nontrivial natural homomorphism into $N^k_{q,k-1}$, which is abelian. Clearly, many homomorphisms from G to $\mathbb{R}^q$ can be constructed by following the natural homomorphism by a homomorphism from $N^k_{q,k-1}$ to $\mathbb{R}^q$. Any element of G with a higher order of vanishing maps into 0, a feature which also appears in the proofs of the lemma, since they compare all elements of the group to those that vanish most slowly. There are examples of leaves which have nontrivial holonomy that is C^∞ tangent to the identity (Cantwell and Conlon [1980, § 5]), so the arbitrariness of the construction is essential. □

The next theorem asserts that in many cases, commuting C^2 germs on the line must behave in more or less the same way. To state the theorem, a preliminary definition is required.

5.3 Definition. A local diffeomorphism f leaving x fixed is a *local contraction* at x if there exists a neighborhood V of x such that $\bigcap_{n \geq 0} f^n(V) = \{x\}$. It is a *local expansion* if its inverse is a local contraction.

5.4 Theorem. *If f and g are commuting orientation preserving C^2 local diffeomorphisms at 0 in $\mathbb{R}$, and f is a local contraction, then g is either the identity, a local contraction, or a local expansion. If f and g are C^k, $k \geq 2$, and f is C^{k-1} tangent to the identity at 0 but not C^k tangent to the identity, then g is completely determined by the value of its k-th derivative at 0.*

Proof. To prove the first statement, it suffices to show that if g has a fixed point other than 0, then g is the identity. If f is linear, then g is linear and the conclusion follows. If $f'(0) \neq 1$ or $g'(0) \neq 1$, then it is C^1 equivalent to a linear map by a theorem of Sternberg [1957], and the conclusion follows from the properties of linear maps. If $f'(0) = g'(0) = 1$, then by commutativity

$$g'(f^k(x)) = g'(x) \prod_{j=0}^{k-1} (f'(f^j \circ g(x))/f'(f^j(x))).$$

Letting k go to infinity yields a differential equation

$$1 = A(x, g(x))\, g'(x)$$

provided the infinite product converges. In fact, it does, and the differential equation has good enough uniqueness properties to yield the desired conclusion. For the details, and the proof of the second statement, see Kopell [1970, Lemma 1]. □

The hypothesis of C^2 is essential. Indeed, Sternberg [1957] has shown that the C^1 map

$$f_a(x) = \begin{cases} ax(1 - (\log|x|)^{-1}) & x \neq 0,\ a > 0,\ |x| < 1 \\ 0 & x = 0 \end{cases}$$

cannot be linearized by a C^1 change of coordinates.

The next result to be discussed is the fixed point theorem of Sacksteder. This theorem guarantees the existence of leaves with nontrivial holonomy if the minimal sets of a C^2 codimension 1 foliation are sufficiently complicated. The statement of the theorem given here is due to Cantwell and Conlon [1980].

5.5 Theorem. *Let N be a union of intervals (open, half-open, or closed) and Γ a pseudogroup on N with a finite generating system Γ_0 composed of C^2 orientation preserving diffeomorphisms with domain a finite union of intervals. Let C be a Γ-invariant subset of the interior of N which is compact, nowhere dense, and equal to its own derived set. Suppose that the interior of the domain of each element of Γ_0 meets C in a compact set, and that q is an endpoint of a gap in C which is a cluster point of its own orbit. Then there is a compact, connected neighborhood U of q in N and a sequence $\{g_m\}$ of elements of Γ whose domains contain U, such that $g_m(q)$ converges to q and $g_m'(t)$ converges to 0 uniformly on U. Furthermore, if V is any compact neighborhood of q contained in U, then for m sufficiently large, g_m maps V into itself and has a unique fixed point x_m in $V \cap C$ such that at x_m, g_m is a 2-sided contraction with derivative less than 1. Also, if $\{q_n\}$ is any sequence in N converging to q, then for n sufficiently large, q belongs to the closure of the orbit of q_n.*

Proof. This is a generalization of the theorem of Sacksteder [1965, Theorem 1] but it follows from the same proof. □

It is possible for a C^∞ foliation of codimension 1 to have a minimal set which is a Cantor set. The first example of this phenomenon is due to Sacksteder [1964]. The total space is the product of a circle with a surface of genus 2, and the foliation is transverse to the circles. The fact that the fundamental groups of many leaves are nonabelian is related both to Theorem 5.4 and to the transverse invariant measures discussed in § 2 and in Theorem 5.7 below. By Theorem 5.5, there must be many leaves with nontrivial holonomy groups.

By contrast, the simplest examples of transversally oriented codimension 1 foliations are those defined by C^∞ closed 1-forms. These foliations are Riemannian, and the holonomy group of each leaf is trivial. If the manifold M is compact, the closed form cannot be exact, since the gradient of a function vanishes at a maximum point. In fact, its cohomology class can be approximated arbitrarily closely by the class pulled back from the fundamental class of the circle by a fibering of M over the circle (Tischler [1970]). Geometrically, this means that the tangent spaces to the fibers can be taken arbitrarily close to the tangent spaces to the leaves. If the foliation is only C^1, then it may have trivial holonomy without being Riemannian. Indeed, as has been remarked in Example I.6.3, there are C^1 1-dimensional foliations of the 2-torus with trivial holonomy which cannot be Riemannian because the minimal set is a Cantor set. On the other hand, C^2

foliations without holonomy of the 2-torus are necessarily topologically equivalent to a Riemannian foliation. Since in codimension 1, being Riemannian is the same as being defined by a closed form, the following Proposition and Theorem generalize this fact to arbitrary dimensions.

5.6 Proposition. *Suppose a compact manifold admits a tranversally oriented C^k foliation of codimension* 1 *having trivial holonomy groups. Suppose further that either* (i) $k \geqq 2$ *or* (ii) *the foliation admits a transverse flow and every leaf is either closed or somewhere dense. The there is a transverse topological flow which preserves the foliation.*

Proof. Consider the transverse flow lines through points x and y on a leaf. Then any path from x to y lying in the leaf defines a holonomy map from some interval on one flow line to an interval on the other flow line. If the maximal interval on which the holonomy map is defined has an endpoint, then the leaf through this endpoint has locally infinite holonomy pseudogroup. By Proposition I.5.10, this contradicts the triviality of the holonomy groups. Hence, each holonomy map is a homeomorphism from $\mathbb{R}$ onto $\mathbb{R}$. If some flow line is a circle, then a homeomorphism of the circle is obtained by using the work of Novikov (Proposition 5.8 below). Since there exists a flow with at least one flow line a circle, the proof of the theorem is thus reduced to the study of homeomorphisms of the circle, and the theorem of Denjoy [1932] can be applied. This proof is due to Imanishi [1974] and the details can be found in his paper. □

5.7 Theorem. *Any C^k, transversally oriented, foliation of codimension* 1 *of a compact manifold having trivial holonomy groups is topologically equivalent to a foliation defined by C^{k-1} closed form provided $k \geqq 2$.*

Proof. If the linearization map given by Denjoy's theorem and used in the proof of Proposition 5.6 is C^k, the volume element on the circle gives rise to the required form. Since in general this map is C^0, one proceeds by constructing a new differential structure on M such that in this structure the same set of leaves forms a C^k foliation defined by a C^{k-1} closed form. Again, the details are given by Imanishi [1974]. □

This theorem is due to Sacksteder [1965], who gives a very different proof, based on the observation that for a Riemannian foliation there is on any transversal a natural volume element which is invariant by all elements of the holonomy pseudogroup. Conversely, it suffices to show that under suitable hypotheses, a pseudogroup acting on the circle admits an invariant measure. Theorem 5.5 is used to show that these hypotheses are satisfied, and is the only place where the assumption of C^2 differentiability is used. This proof was the beginning of the theory of transverse invariant measures for foliations which is described in § 2.

If a foliation is defined by a closed 1-form, then any positive constant multiple of the form is also a closed form defining the same foliation and the same transverse orientation. Thus, the foliation defines a ray in $H^1(M:\mathbb{R})$. As a consequence of Theorem 5.7, this ray is well-defined as soon as the foliation is C^2 and without holonomy on a compact manifold (Roussarie [1973, p. 129]). On the product of a circle with a compact oriented 2-manifold, this ray classifies such foliations up to topological isotopy (Roussarie [1973]). On the 3-torus, two such foliations are topologically equivalent if and only if the corresponding rays are rationally equivalent (Lokot' [1972]). The techniques of Roussarie have been extended by Moussu and Roussarie [1973] to 3-dimensional manifolds foliated with the boundary as a union of leaves and no holonomy in the interior. It then follows from Theorem 5.4 that the holonomy on the boundary is abelian, and from this a classification up to isotopy is derived.

If a foliation is defined by a C^1 closed 1-form, then it is a C^2 foliation, so the Godbillon-Vey class is defined and is in fact 0. This conclusion has been extended to C^2 foliations without holonomy by Morita and Tsuboi [1980].

An important early result about foliations without holonomy is due to Novikov [1965]. In the proof he needs normal vector fields with unique trajectories. For a C^2 foliation, these vector fields can be obtained from the associated C^1 $(n-1)$-plane field and an arbitrary C^1 Riemannian metric. Thus, the following proposition is true for C^2 foliations, though it was originally stated for C^∞ foliations.

5.8 Proposition. *Let M be compact and $\mathscr{F}$ be a C^k foliation, $k \geqq 2$, transversally oriented and without holonomy. Then the universal covering space $\tilde{M}$ is C^{k-1} diffeomorphic to $\tilde{L} \times \mathbb{R}$, where $\tilde{L}$ is the universal cover of a leaf L of $\mathscr{F}$ and each $\tilde{L} \times \{t\}$ is a leaf of the induced foliation $\tilde{\mathscr{F}}$ of $\tilde{M}$.*

If the foliation has holonomy, the methods used in the proofs of Theorems 5.4 and 5.5 can be modified to produce much information about the structure of the closure of a leaf.

5.9 Proposition. *Let $\mathscr{F}$ be a transversally oriented C^2 foliation of codimension 1 of a compact oriented manifold M, and let U be a saturated open subset of M. Then for any leaf L contained in U, $\bar{L} \cap U$ contains a minimal set of the restriction of $\mathscr{F}$ to U.*

This result is obtained by studying the action of the holonomy pseudogroup on a family of short transversals at points of $(\bar{L} - L) \cap U$. Proofs have been given by Hector [1972] and Cantwell and Conlon [1980]. Since deleting from U a set minimal in U leaves an open set, the process can be repeated and a hierarchiy of local minimal sets established. If this process does not exhaust $\bar{L}$, then the complement is an uncountable union of leaves, each of which has trivial holonomy jets, but approaches itself from both sides. For the further study of local minimal sets, the notion of end of a manifold is useful. An end is, intuitively, a particular direction for going to infinity on a noncompact manifold. The set ε of ends can be topologized as a compact, totally disconnected metrizable space, and thus has a family ε^γ of derived sets indexed by ordinals γ. For a leaf, an end is a possible direction for other leaves to wind into it. Given a leaf L of nonexponential growth (Definition 2.22), each local minimal set in $\bar{L}$ consists of leaves of polynomial growth of degree k such that ε^k is empty – this k is also the level of the minimal set in its hierarchy. If all leaves have nonexponential growth and level at most 1, then the Godbillon-Vey class vanishes. (It is conjectured that the restriction on levels is unnecessary.) For proofs, related results, and other references, see Cantwell and Conlon [1981].

For the study of foliations of codimension 1 on 3-dimensional manifolds, the fundamental result is due to Novikov [1965].

5.10 Theorem. *Let $\mathscr{F}$ be a C^2 foliation of a compact 3-dimensional manifold M. Suppose that f: $S^1 \times I \to M$ is a C^1 mapping such that if $f_t(x) = f(x, t)$, then*

(i) *f_t is an embedding of S^1 into a leaf $L(t)$.*

(ii) *$f_0(S^1)$ is not nullhomotopic in $L(0)$, but for $t > 0$, $f_t(S^1)$ is nullhomotopic in $L(t)$.*

(iii) *For each x, the curve $t \mapsto f(x, t)$ is transverse to $\mathscr{F}$.*

Then $L(0)$ is a torus which bounds a Reeb component whose interior leaves are the leaves $L(t)$ for $t > 0$.

Proof. See Rosenberg and Roussarie [1970], who prove the theorem in this form. The original, slightly more general form may be incorrect, since a counterexample to an important step in the proof has been given by Muller [1980]. The hypothesis of C^2 is required for the same reasons as in the proof of Proposition 5.8. □

5.11 Corollary. *Any foliation of a compact 3-dimensional manifold with finite fundamental group has a compact leaf which is either a torus or a Klein bottle.*

Proof. The universal covering space is compact, and on it all the induced data are orientable. By an argument of Haefliger [1958, 1968], there exists a closed curve on some leaf which is not nullhomotopic on the leaf, but which is nullhomotopic on all nearby leaves on one side. Thus, Theorem 5.10 can be applied to prove the existence of a toral leaf. Hence, on M there is a leaf covered by a torus. □

A number of other results about 2-dimensional foliations of 3-manifolds can be obtained as applications of Theorem 5.10. Novikov [1965] shows that if $\pi_2(M) \neq 0$, but $\pi_2(L) = 0$ for every leaf L, then there is necessarily a Reeb component. By the contrapositive, if there are no compact leaves, then the universal cover is contractible. Rosenberg and Roussarie [1970] classify foliations such that all leaves are either planes or tori.

Bibliography

Abe, K.

1971 Characterization of totally geodesic submanifolds of S^N and CP^N by an inequality. Tôhoku Math. J. (2) *23*, 219–244.

Anosov, D. V.

1962 Roughness of geodesic flows on compact Riemannian manifolds of negative curvature. Dokl. Akad. Nauk SSSR 145, 707–709 [Russian]. Translation: Soviet Math. Dokl. *3*, 1068–1069.

1963 Ergodic properties of geodesic flows on closed Riemannian manifolds of negative curvature. Dokl. Akad. Nauk SSSR *151*, 1250–1252 [Russian]. Translation: Soviet Math. Dokl. *4*, 1153–1156.

1967 Geodesic flows on closed Riemann manifolds with negative curvature. Trudy Mat. Inst. Steklov *90* [Russian]. Translation: Proc. Steklov Inst. Math. *90*.

Arnold, V. I.

1973 Ordinary Differential Equations. Cambridge, MA-London: M.I.T.

Artin, E.

1957 Geometric Algebra. New York: Interscience.

Asimov, D.

1978 On the average Gaussian curvature of leaves of foliations. Bull. Amer. Mat. Soc. *84*, 131–133.

Asimov, D. and Gluck, H.

1980 Morse-Smale fields of geodesics. In: Global Theory of Dynamical Systems, pp. 1–17. Lecture Notes in Math., vol. *819*. Berlin-Heidelberg-New York: Springer.

Baum, P. and Bott, R.

1972 Singularities of holomorphic foliations. J. Differential Geometry *7*, 279–342.

Bernshtein, I. N. and Rozenfeld, B. I.

1972 On characteristic classes of foliations. Funkcional. Anal. i Prilozen *6* no. *1*, 68–69 [Russian]. Translation: Functional Anal. Appl. *6*, 60–61.

1973 Homogeneous spaces of infinite-dimensional Lie algebras and characteristic classes of foliations. Uspehi Mat. Nauk *28* no. *4*, 103–138 [Russian]. Translation: Russian Math. Surveys *28* no. *4*, 107–142.

Blumenthal, R. A.

1980 The base-like cohomology of a class of transversely homogeneous foliations. Bull. Sci. Math.(2) *104*, 301–303.

1983 Riemannian foliations with parallel curvature. Nagoya Math.J.

Born, M.

1949 Natural Philosophy of Cause and Chance. Oxford: Clarendon.

Bott, R.

1970 On a topological obstruction to integrability. In: Global Analysis, Proceedings of Symposia in Pure Math., vol. *16*, pp. 127–131. Providence, RI: Amer. Math. Soc.

1972 Lectures on characteristic classes and foliations. In: Lectures on Algebraic and Differential Topology Delivered at the II. ELAM, pp. 1–94. Lecture Notes in Math., vol. *279*. Berlin-Heidelberg-New York: Springer.

1973 On the Chern-Weil homomorphism and the continuous cohomology of Lie groups. Adv. in Math. *11*, 289–303.

Bott, R.
1975 Some aspects of invariant theory in differential geometry. In: Differential Operators on Manifolds, C.I.M.E. 3 Ciclo 1975, pp. 49–145. Roma: Cremonese.
1976 On characteristic classes in the framework of Gelfand-Fuks cohomology. Astérisque 32–33, 113–139.
Bott, R. and Heitsch, J. L.
1972 A remark on the integral cohomology of $B\Gamma_q$. Topology *11*, 141–146.
Bott, R., Shulman, H., and Stasheff, J.
1976 On the deRham theory of certain classifying spaces. Adv. in Math. *20*, 43–56.
Bourbaki, N.
Éléments de Mathématique. II. Algèbre. VII. Groupes et algèbres de Lie. Paris: Hermann.
Brandt, H.
1913 Zur Komposition der quaternären quadratischen Formen. J. Reine Angew. Math. *143*, 106–127.
1927 Über eine Verallgemeinerung des Gruppenbegriffes. Math. Ann. *96*, 360–366.
Brito, F., Langevin, R., and Rosenberg, H.
1977 Intégrales de courbure sur une variété feuilletée. C.R. Acad. Sci. Paris Sér. A–B *285*, A533–A536.
Brown, E. H. Jr.
1965 Abstract homotopy theory. Trans. Amer. Math. Soc. *119*, 79–85.
Buffet, J.-P. and Lor, J.-C.
1970 Une construction d'un universal pour une classe assez large de Γ-structures. C. R. Acad. Sci. Paris Sér. A–B *270*, A640–A642.
Cantwell, J. and Conlon, L.
1981 Poincare-Bendixson theory for leaves of codimension one. Trans. Amer. Soc. *265*, 181–209.
Carathéodory, C.
1909 Untersuchungen über die Grundlagen der Thermodynamik. Math. Ann. *67*, 355–386.
Carriére, Y.
1981 Flots riemanniens et feuilletages géodesibles de codimension un. Thesis, University of Lille.
Cartan, E.
1899 Sur certaines expressions différentielles et le problème de Pfaff. Ann. Sci. École Norm. Sup. (3) *16*, 239–332. Oeuvres II, 303–396.
1901 Sur l'intégration des systèmes d'équations aux différentielles totales. Ann. Sci. École Norm. Sup. (3) *18*, 241–311. Oeuvres II, 411–481.
1904/05 Sur la structure des groupes infinis de transformations. Ann. Sci. École Norm. Sup. (3) *21*, 153–206; *22*, 219–308. Oeuvres II, 571–624, 625–714.
1908 Les sous-groupes des groupes continus de transformations. Ann. Sci. École Norm. Sup. (3) *25*, 57–194. Oeuvres II, 719–856.
1909 Les groupes de transformations continus, infinis, simples. Ann. Sci. École Norm. Sup. (3) *26*, 93–161. Oeuvres II, 857–925.
1945 Les Systèmes Différentielles Extérieurs et leurs Applications Géométriques. Paris: Hermann.
Cartan, H.
1949/50 Cohomologie réelle d'un espace fibré principal différentiable. In: Sem. Cartan 1949/50, exp. 19–20. Paris: École Norm. Sup.
Cartan, H. and Eilenberg, S.
1956 Homological Algebra. Princeton, NJ: Princeton Univ.
Chern, S. S.
1953 Pseudo-groupes continus infinis. In: Colloq. Inter. C.N.R.S., Géométrie Différentielle, Strasbourg, pp. 119–136. Paris: Centre National de Recherche Scientifique.
1955 An elementary proof of the existence of isothermal parameters on a suface. Proc. Amer. Math. Soc. *6*, 771–782.
Chern, S. S. and Simons, J.
1974 Characteristic forms and geometric invariants. Ann. of Math (2) *99*, 48–69.

Choquet, G.
1956 Existence et unicité des representations intégrales au moyen des points extrémaux dans les cônes convexes. In: Sém. Bourbaki 1956/57, exp. 139. New York: W. A. Benjamin.
1969 Lectures on Analysis, Vol II, Representation Theory. Reading, MA: Benjamin/Cummings.
Chow, W. L.
1940/41 Über Systeme von linearen partiellen Differentialgleichungen erster Ordnung. Math. Ann. *117*, 98–105.
Clebsch, A.
1866 Über die simultane Integration linearer partieller Differentialgleichungen. J. Reine Angew. Math. *65*, 257–268.
Connes, A.
1979 Sur la théorie noncommutative de l'intégration. In: Algèbres d'Opérateurs, pp. 19–143. Lecture Notes in Math., vol. 725. Berlin-Heidelberg-New York: Springer.
1980 Feuilletages et algèbres d'opérateurs. In: Sém. Bourbaki 1979/80, exp. 551. Lecture Notes in Math., vol. *842*. Berlin-Heidelberg-New York: Springer.
1981 A survey of foliations and operator algebras. I.H.E.S.
Deahna, F.
1840 Über die Bedingungen der Integrabilität linearer Differentialgleichungen erster Ordnung zwischen einer beliebigen Anzahl veränderlicher Größen. J. Reine Angew. Math. *20*, 340–349.
Denjoy, A.
1932 Sur les courbes définis par les équations différentielles à la surface du tore. J. Math. Pures Appl. (9) *11*, 333–375.
Dieudonné, J.
Éléments d'Analyse. Paris: Gauthier-Villars.
Dombrowski, P.
1978 Jacobi fields, totally geodesic foliations, and geodesic differential forms. Resultate Math. *1*, 156–194.
Edwards, R., Millett, K., and Sullivan, D.
1977 Foliations with all leaves compact. Topology *16*, 13–32.
Ehresmann, C.
1951 Les prolongements d'une variété différentiable. I. Calcul des jets. II. L'espace des jets d'ordre r de V_n dans V_m. III. Transitivité des prolongements. C. R. Acad. Sci. Paris *233*, 598–600, 777–779, 1081–1083.
1952a Structures locales et structures infinitesimales. C. R. Acad. Sci. Paris *243*, 587–589.
1952b Les prolongements d'une variété différentiable. IV. Éléments de contact et éléments d'envelope. V. Covariants différentiels et prolongements d'une structure infinitésimale. C. R. Acad. Sci. Paris *234*, 1028–1030, 1424–1425.
Ehresmann, C. and Reeb, G.
1944 Sur les champs d'éléments de contact de dimension p complètement intégrables dans une variété continuement différentiable. C. R. Acad. Sci. Paris *218*, 955–957.
Epstein, D. B. A.
1976 Foliations with all leaves compact. Ann. Inst. Fourier (Grenoble) *26*, 265–282.
1977 A topology for the space of foliations. In: Geometry and Topology, Proc. III. Latin Amer. School of Math., pp. 132–150. Lecture Notes in Math., vol. *597*. Berlin-Heidelberg-New-York: Springer.
Epstein, D. B. A., Millett, K. C., and Tischler, D.
1977 Leaves without holonomy. J. London Math. Soc. (2) *16*, 548–552.
Epstein, D. B. A. and Vogt, E.
1978 A counterexample to the periodic orbit conjecture in codimension 3. Ann. of Math. (2) *108*, 539–552.
van Est, W. T.
1953 Group cohomology and Lie algebra cohomology in Lie groups. Nederl. Akad. Wetensch. Indag. Math. *15*, 484–492, 493–504.
1955a On the algebraic cohomology concepts in Lie groups. Nederl. Akad. Wetensch. Indag. Math. *17*, 225–233, 286–294.

van Est, W. T.
1955b Une application d'une méthode de Cartan-Leray. Nederl. Akad. Wetensch. Indag. Math. *17*, 542–544.
1958 A generalization of the Cartan-Leray spectral sequence. Nederl. Akad. Wetensch. Indag. Math. *20*, 399–413.
Fedida, E.
1971 Sur les feuilletages de Lie. C. R. Acad. Sci. Paris Sér. A–B *272*, A999–A1002. Thesis, University of Strasbourg.
Ferus, D.
1970 Totally geodesic foliations. Math. Ann. *188*, 313–316.
Forsyth, A. R.
1890 Theory of Differential Equations, First Part. Cambridge: University.
Frobenius, F. G.
1877 Über das Pfaffsche Problem. J. Reine Angew. Math. *82*, 267–282. Ges. Abh. I, 286–301.
Fuks, D.
1978 Cohomology of infinite-dimensional Lie algebras and characteristic classes of foliations. Itogi Nauki-Serija "Matematika" *10*, 179–286 [Russian]. Translation: J. Soviet Math. *11* (1979), 922–980.
Galois, E.
1846 Oeuvres mathématiques d'Evariste Galois. J. Math. Pures Appl. *11*, 381–444.
Gelfand, I. M. and Fuks, D. B.
1970 The cohomology of the Lie algebra of formal vector fields. Izv. Akad. Nauk SSSR *34*, 322–337 [Russian]. Math. USSR - Izv. *4*, 327–342.
Gelfand, I. M., Kalinin, D. I., and Fuks, D. B.
1972 On the cohomology groups of the Lie algebra of Hamiltonian formal vector fields. Funkcional. Anal. i Priložen *6*, no. *3*, 25–29 [Russian]. Translation: Functional Anal. Appl. *6*, 193–196.
Gelfand, I. M., Kazhdan, D. A., and Fuks, D. B.
1972 The actions of infinite dimensional Lie algebras. Funkcional. Anal. i Priložen *6*, no. *1*, 10–15 [Russian], Functional Anal. Appl. *6*, 9–13.
Girbau, J. and Nicolau, M.
1980 Pseudo-differential operators on V-manifolds and foliations II. Collect. Math. *31*, 63–95.
Gluck, H.
1980 Dynamical behaviour of geodesic fields. In: Global Theory of Dynamical Systems, pp. 190–215. Lecture Notes in Math., vol. *819*. Berlin-Heidelberg-New York: Springer.
Gluck, H. and Warner, F. W.
1983 Great circle fibrations of the three sphere. Duke Math. J. *50*.
Godbillon, C.
1972 Cohomologies d'algèbres de Lie de champs de vecteurs formels. In: Sém. Bourbaki 1972/73, exp. 421. Lecture Notes in Math., vol. *383*. Berlin, Heidelberg-New York: Springer.
Godbillon, C. and Vey, J.
1971 Un invariant des feuilletages de codimension un, C. R. Acad. Sci. Paris Sér. A–B *273*, A92–A95.
Greenleaf, F. P.
1969 Invariant Means on Topological Groups and their Applications. Van Nostrand Math. Studies, vol. *16*. New York-Toronto-London: Van Nostrand-Reinhold.
Gromov, M. L.
1968 Transversal mappings of foliations, Dokl. Akad. Nauk SSSR *182*, no. *2*, 255–258 [Russian]. Translation: Soviet Math. Dokl. *9*, 1126–1129.
1969 Stable mappings of foliations into manifolds. Izv. Akad. Nauk SSSR Ser. Mat. *33*, 707–734. Translation: Math. USSR - Izv. *3*, 671–694.
Grothendieck, A.
1951 Sur une notion de produit tensoriel topologique d'espaces vectoriels topologiques, et une classe remarquable d'espaces vectoriels liée à cette notion. C. R. Acad. Sci. Paris *233*, 1556–1558.

Guillemin, V.
1973 Cohomology of vector fields on a manifold. Adv. in Math. *10*, 192–220.
Guillemin, V. and Sternberg, S.
1966 Deformation theory of pseudogroup structures. Mem. Amer. Math. Soc. *64*.
Haefliger, A.
1956 Sur les feuilletages analytiques. C. R. Acad. Sci. Paris *242*, 2908–2910.
1958 Structures feuilletées et cohomologie à valeur dans un faisceau de groupoïdes. Comment. Math. Helv. *32*, 248–329.
1962 Variétés feuilletées. Ann. Scuola Norm. Sup. Pisa Sci. Fis. Mat. (3) *16*, 367–397.
1968 Travaux de Novikov sur les feuilletages. In: Sém. Bourbaki 1967/68, exp. 339. New York: W. A. Benjamin.
1970 Feuilletages sur les variétés ouvertes. Topology *9*, 183–194.
1971 Homotopy and integrability. In: Manifolds – Amsterdam 1970, pp. 133–163. Lecture Notes in Math., vol. *197*. Berlin-Heidelberg-New York: Springer.
1972 Sur les classes caractéristiques des feuilletages. In: Sém. Bourbaki 1971/72, exp. 412. Lectures Notes in Math., vol. *317*. Berlin-Heidelberg-New York: Springer.
1976 Sur la cohomologie de l'algèbre de Lie des champs de vecteurs. Ann. Sci. École Norm. Sup. (4) *9*, 503–532.
1980 Some remarks on foliations with minimal leaves. J. Differential Geom. *15*, 269–284.
Hamilton, R. S.
1982 The inverse function theorem of Nash and Moser. Bull. Amer. Math. Soc. (N.S.) *7*, 65–222.
Harrison, J.
1975 Unsmoothable diffeomorphisms. Ann. of Math. (2) *102*, 83–94.
1979 Unsmoothable diffeomorphisms on higher dimensional manifolds. Proc. Amer. Math. Soc. *73*, 249–255.
Hector, G.
1972 Ouverts incompressibles et théorème de Denjoy-Poincaré pour les feuilletages. C. R. Acad. Sci. Paris Sér. A–B *274*, A159–A162. Thesis, University of Strasbourg.
1977 Feuilletages en cylindres. In: Geometry and Topology, Proc. III. Latin Amer. School of Math., pp. 252–270. Lecture Notes in Math., vol. *597*. Berlin-Heidelberg-New York: Springer.
Heitsch, J. L.
1973 Deformations of secondary characteristic classes. Topology *12*, 381–388.
1978 Independent variation of secondary classes. Ann. of Math. (2) *108*, 421–460.
Herman, M. R.
1976a Conjugaison C^∞ des difféomorphismes du cercle dont le nombre de rotation satisfait à une condition arithmétique. C. R. Acad. Sci. Paris Sér. A–B 282, A503–A506.
1976b Conjugaison C^∞ des difféomorphismes du cercle pour presque tout nombre de rotation. C. R. Acad. Sci. Paris Sér. A–B, A579–A582.
1979 Sur la conjugaison différentiable des difféomorphismes du cerle à des rotations. Inst. Hautes Études Sci. Publ. Math. *49*, 5–233.
Hermann, R.
1960a A sufficient condition that a mapping of Riemannian manifolds be a fiber bundle. Proc. Amer. Math. Soc. *11*, 236–242.
1960b The differential geometry of foliations I. Ann. of Math. (2), *72*, 445–457.
1962 The differential geometry of foliations II. J. Math. Mech. *11*, 305–315.
1979 The Born theory of rigid motions in relativity and the theory of Riemannian foliations. In: Development of Mathematics in the Nineteenth Century, Appendices, Kleinian Mathematics from an Advanced Standpoint, pp. 549–573. Brookline, MA: Math Sci.
Hirsch, M. W. and Thurston, W.
1975 Foliated bundles, invariant measures, and flat manifolds. Ann. of Math. (2) *101*, 369–390.
Hörmander, L.
1964 The Frobenius-Nirenberg theorem. Ark. Mat. *5*, 425–432.

Hurder, S.
1981a On the homotopy and cohomology of the classifying space of Riemannian foliations. Proc. Amer. Math. Soc. *81*, 485–489.
1981b Dual homotopy invariants of G-foliations. Topology *20*, 365–387.
1981c On the secondary classes of foliations with trivial normal bundles. Comment. Math. Helv. *56*, 307–326.
Husemoller, D.
1966 Fiber Bundles. New York-Toronto-London-Sydney: McGraw-Hill.
1975 2nd. Ed. Berlin-Heidelberg-New York: Springer.
Imanishi, H.
1974 On the theorem of Denjoy-Sacksteder for codimension one foliations without holonomy. J. Math. Kyoto Univ. *14*, 607–634.
Johnson, D. L. and Naveira, A. M.
1981 A topological obstruction to the geodesibility of a foliation of odd dimension. Geom. Dedicata *11*, 347–352.
Kähler, E.
1934 Einführung in die Theorie der Systeme von Differentialgleichungen. Leipzig: Teubner.
Kamber, F. and Tondeur, P.
1971 Invariant differential operators and the cohomology of Lie algebra sheaves. Mem. Amer. Math. Soc. *113*.
1973 Cohomologie des algèbres de Weil relatives tronquées. Algèbres de Weil semisempliciales. Homomorphisme charactéristique d'un fibré principal feuilleté. Classes charactéristiques dérivées d'un fibré principal feuilleté. C. R. Acad. Sci. Paris Sér. A–B *276*, A459–A462, A1177–A1179, A1407–1410, A1449–1452.
1974 Characteristic invariants of foliated bundles. Manuscripta Math. *11*, 51–89.
1975 Foliated Bundles and Characteristic Classes. Lecture Notes in Math., vol. *493*. Berlin-Heidelberg-New York: Springer.
1982 Harmonic foliations. In: Proc. NSF Conference on Harmonic Maps, Tulane (Dec. 1980), pp. 87–121. Lecture Notes in Math., vol. *949*. Berlin-Heidelberg-New York: Springer.
van Kampen, E. R.
1935 The topological transformations of a simple closed curve into itself. Amer. J. Math. *57*, 141–152.
Kitahara, H.
1979 Remarks on square-integrable basic cohomology spaces on a foliated Riemannian manifold. Kodai Math. J. *2*, 187–193.
Klein, F.
1872 Vergleichende Betrachtungen über neuere geometrische Forschungen. Erlangen: Andreas Deichert.
1893 Math. Ann. *43*, 63–100 (with additional notes).
Kobayashi, S.
1961 Canonical forms on frame bundles of higher order contact. In: Differential Geometry, Proc. of Symposia in Pure Math., vol. *3*, pp. 186–193. Providence, RI: Amer. Math. Soc.
1972 Transformation Groups in Differential Geometry. Ergebnisse Math. Grenzgeb., vol. 70. Berlin-Heidelberg-New York: Springer.
Kobayashi, S. and Nomizu, K.
1963 Foundations of Differential Geometry, Vol. I. New York-London: Interscience.
1969 Vol. II
Kopell, N.
1970 Commuting diffeomorphisms. In: Global Analysis, Proc. of Symposia in Pure Math., vol. *14*, pp. 165–184. Providence, RI: Amer. Math. Soc.
Koszul, J.-L.
1950 Homologie et cohomologie des algèbres de Lie. Bull. Soc. Math. France *78*, 65–128.
Kumpera, A. and Spencer, D. C.
1972 Lie Equations. Volume I: General Theory. Ann. of Math. Studies, vol. *73*. Princeton, NJ: Princeton Univ.

Kupka, I.
1964 The singularities of integrable structurally stable Pfaffian forms. Proc. Nat. Acad, Sci. USA. *52*, 1431–1432.

Lickorish, W. B. R.
1965 A foliation for 3-manifolds, Ann. of Math. (2) *82*, 414–420.

Lichnerowicz, A.
1976 Sur l'algèbre de Lie des champs de vecteurs. Comment. Math. Helv. *51*, 343–368.
1977 Les variétés de Poisson et leurs algèbres de Lie associées. J. Differential Geom. *12*, 253–300.

Lokot', T. V.
1972 Topological classification of foliations of codimension 1 with trivial holonomy group of class C^2 on the 3-dimensional torus. Uspehi Mat. Nauk *27*, no. *3*, 205 [Russian].

Malgrange, B.
1976 Frobenius avec singularites I. Codimension un. Inst. Hautes Études Sci. Publ. Math. *46*, 163–173.
1977 2. Le cas général. Invent. Math. *39*, 67–89.

Milnor, J.
1956 Construction of universal bundles II. Ann. of Math. (2) *63*, 430–436.

Molino, P.
1971 Classe d'Atiyah d'un feuilletage et connexions transverses projetables. Classes caractéristiques et obstruction d'Atiyah pour les fibrés principaux feuilletés. C. R. Acad. Sci. Paris Sér. A–B *272*, A779–A781, A1376–1378.
1972 Feuilletages et classes caractéristiques. Symposia Mathematica *10*, 199–209.
1973 Propriétés cohomologiques et propriétés topologiques des feuilletages à connexions transverses projetables. Topology *12*, 317–325.
1977 Feuilletages transversalement complets et applications. Ann. Sci. École Norm. Sup. (4) *10*, 289–307.

Montesinos, A.
1983 On certain classes of almost product structures. Michigan Math. J. *30*.
1982 Conformal curvature for the normal bundle of a conformal foliation. Ann. Inst. Fourier (Grenoble) *32*, no. 3, 261–274.

Morita, S. and Tsuboi, T.
1980 The Godbillon-Vev class of codimension one foliations without holonomy. Topology *19*, 43–49.

Moser, J.
1965 On the volume elements on a manifold. Trans. Amer. Math. Soc. *120*, 286–294.

Mostow, M.
1976 Continuous cohomology of spaces with two topologies. Mem. Amer. Math. Soc. *175*.

Moussu, R.
1976 Sur l'existence d'intégrales premières pour un germe de forme de Pfaff. Ann. Inst. Fourier (Grenoble) *26*, no. *2*, 171–220.

Moussu, R. and Roussarie, R.
1974 Relations de conjugaison et de cobordisme entre certains feuilletages, Inst. Hautes Études Sci. Publ. Math. *43*, 143–168.

Muller, M.-P.
1980 Sur les composantes de Novikov des feuilletages. Topology *19*, 199–201.

Naveira, A. M.
1982 A classification of Riemannian almost product manifolds. Rend. Math. (7)

Nemytskii, V. V. and Stepanov, V. V.
1960 Qualitative Theory of Differential Equations. Princeton, NJ: Princeton Univ.

Newlander, A. and Nirenberg, L.
1957 Complex analytic coordinates in almost complex manifolds. Ann. of Math. (2) *65*, 391–404.

Nijenhuis, A.
1952 Theory of the Geometric Object. Thesis, University of Amsterdam.

Nijenhuis, A. and Woolf, W. B.
1963 Some integration problems in almost-complex and complex manifolds. Ann. of Math. (2) *77*, 424–489.

Nirenberg, L.
1958 A complex Frobenius theorem. In: Conference on Analytic Functions, pp. 172–189. Princeton, NJ: Inst. Advanced Study.

Nishikawa, S. and Sato, H.
1976 On characteristic classes of Riemannian, conformal, and projective foliations. J. Math. Soc. Japan *28*, 223–241.

Novikov, S. P.
1965 Topology of foliations. Trudy Moskov. Mat. Obšč. *14*, 248–278 [Russian].

O'Neill, B.
1966 The fundamental equations of a submersion. Michigan Math. J. *13*, 459–469.

Phillips, A.
1968 Foliations of open manifolds. I. Comment. Math. Helv. *43*, 204–211.
1969 II. Comment. Math. Helv. *44*, 367–370.
1970 Smooth maps transverse to a foliation. Bull. Amer. Math. Soc. *76*, 792–797.

Plante, J. F.
1975 Foliations with measure preserving holonomy. Ann. of Math. *102*, 327–361.

Poincaré, H.
1921 Rapport sur les travaux de M. Cartan. Acta Math. *38*, 137–145.

Reeb, G.
1947 Variétés feuilletées, feuilles voisines. C. R. Acad. Sci. Paris *224*, 1613–1614.
1952 Sur Certaines Propriétés Topologiques des Variétés Feuilletées. Paris: Hermann.

Reeb, G. and Schweitzer, P. A.
1978 Un théorème de Thurston établi au moyen de l'analyse non standard. In: Differential Topology, Foliations, and Gelfand-Fuks Cohomology, p. 138. Lecture Notes in Math., vol. *652*. Berlin-Heidelberg-New York: Springer.

Reinhart, B. L.
1958 Harmonic integrals on almost product manifolds. Trans. Amer. Math. Soc. *88*, 243–276.
1959a Foliated manifolds with bundle-like metrics. Ann. of Math. (2) *69*, 119–132.
1959b Harmonic integrals on foliated manifolds. Amer. J. Math. *81*, 529–536.
1959c Line elements on the torus. Amer. J. Math. *81*, 617–631.
1961 Closed metric foliations. Michigan Math. J. *8*, 7–9.
1971 Automorphisms and integrability of plane fields. J. Differential Geom. *6*, 263–266.
1977 The second fundamental form of a plane field. J. Differential Geom. *12*, 619–627.
1982 Some remarks on the structure of the Lie algebra of formal vector fields. Astérisque.

de Rham, G.
1955 Variétés Différentiables. Paris: Hermann.

Rosenberg, H. and Roussarie, R.
1970 Reeb foliations. Ann. of Math. (2) *91*, 1–24.

Roussarie, R.
1973 Plongement dans les variétés feuilletées et classification de feuilletages sans holonomie. Inst. Hautes Études Sci. Publ. Math. *43*, 101–104.

Rummler, H.
1979 Quelques notions simples en géométrie riemannienne et leurs applications aux feuilletages compacts. Comment. Math. Helv. *54*, 224–239.

Sacksteder, R.
1964 On the existence of exceptional leaves in foliations of codimension one. Ann. Inst. Fourier (Grenoble) *14*, 221–226.
1965 Foliations and pseudogroups. Amer. J. Math. *87*, 79–102.

Sacksteder, R. and Schwartz, A. J.
1965 Limit sets of foliations. Ann. Inst. Fourier (Grenoble) *15*, 201–214.

Schachermeyer, W.
1978 Addendum: Une modification standard de la démonstration non standard de Reeb et

Schweitzer. In: Differential Topology, Foliations, and Gelfand-Fuks Cohomology, pp. 139–140. Lecture Notes in Math., vol. *652*. Berlin-Heidelberg-New York: Springer.

Schwartz, A. J.
1963 A generalization of the Poincaré-Bendixson theorem to closed two-dimensional manifolds. Amer. J. Math. *85*, 453–458.

Schwartz, L.
1957 Théorie des Distributions. Paris: Hermann.
1966 2nd. Edition.

Schwartzman, S.
1957 Asymptotic cycles. Ann. of Math. *66*, 270–284.

Schwarz, G.
1974 On the de Rham cohomology of the leaf space of a foliation. Topology *13*, 185–188.

Schweitzer, P. A.
1978 Editor: Differential Topology, Foliations, and Gelfand-Fuks Cohomology. Lecture Notes in Math., vol. *652*. Berlin-Heidelberg-New York: Springer.

Siegel, C. L.
1945 Note on differential equations on the torus. Ann. of Math. (2) *46*, 423–428.

Stefan, P.
1974 Accessible sets, orbits, and foliations with singularities. Proc. London Math. Soc. (3) *29*, 699–713.

Sternberg, S.
1957 Local C^n transformations of the real line. Duke Math. J. *24*, 97–102.
1964 Lectures on Differential Geometry. Englewood, NJ: Prentice-Hall.

Stiefel, E.
1936 Richtungsfelder und Fernparallelismus in n-dimensionalen Mannigfaltigkeiten. Comment. Math. Helv. *8*, 305–353.

Sullivan, D.
1976a Cycles for the dynamical study of foliated manifolds and complex manifolds. Invent. Math. *36*, 225–255.
1976b A new flow. Bull. Amer. Math. Soc. *82*, 331–332.
1977 Infinitesimal computations in topology. Inst. Hautes Études Sci. Publ. Math. *47*, 269–331.
1979 A homological characterization of foliations consisting of minimal surfaces. Comment. Math. Helv. *54*, 218–223.

Sullivan, D. and Williams, R.
1976 Homology of attractors. Topology *15*, 259–262.

Sussmann, H. J.
1973 Orbits of families of vector fields and integrability of distributions. Trans. Amer. Math. Soc. *180*, 171-188.

Terng, C. L.
1978 Natural vector bundles and natural differential operators. Amer. J. Math. *100*, 775–828.

Thom, R.
1954 Quelques propriétés globales des variétés différentiables. Comment. Math. Helv. *28*, 17–86.
1964 Généralisation de la théorie de Morse aux variétés feuilletées. Ann. Inst. Fourier (Grenoble) *14*, 173–190.
1975 On singularities of foliations. In: Manifolds Tokyo 1973, pp. 171–173. Tokyo: Univ. of Tokyo.

Thomas, J. M.
1937 Differential Systems. Colloquium Publication 21. Providence, RI: Amer. Math. Soc.

Thurston, W.
1972 Noncobordant foliations of S^3. Bull. Amer. Math. Soc. *78*, 511–514.
1974a The theory of foliations of codimension greater than one. Comment. Math. Helv. *49*, 214–231.
1974b A generalization of the Reeb stability theorem. Topology *13*, 347–352.

1974c Foliations and groups of diffeomorphisms. Bull. Amer. Math. Soc. *80*, 304–307.
1976 Existence of codimension-one foliations. Ann. of Math. (2) *102*, 249–268.

Tischler, D.
1970 On fibering certain foliated manifolds over S^1. Topology *9*, 153–154.

Trauber, P.
1973 The continuous cohomology of the Lie algebra of vector fields on a smooth manifold. Thesis, Princeton University.

Vaisman, I.
1971 Variétés riemanniennes feuilletées. Czechoslovak Math. J. *21*, 46–75.
1974 Remarks about differential operators on foliate manifolds. An. Şti. Univ. "Al. I. Cuza" Iaşi Secţ. Ia Mat. *20*, 327–350.
1979 Conformal foliations. Kodai Math. J. *2*, 26–37.

Veblen, O. and Whitehead, J. H. C.
1932 The Foundations of Differential Geometry. Cambridge: University.

Vidal, E.
1964 Sobre algunos problemas en relación con la medida en espacios foliados. In: I Coloquio Internacional de Geometría Diferencial, pp. 63–77. Santiago: Universidad.
1967 On regular foliations. Ann. Inst. Fourier (Grenoble) *17*, 129–133.

Vogt, E.
1976 Foliations of codimension 2 with all leaves compact. Manuscripta Math. *18*, 187–212.
1977 A periodic flow with infinite Epstein hierarchy. Manuscripta Math. *22*, 403–412.

Wadsley, A. W.
1975 Geodesic foliations by circles. J. Differential Geom. *10*, 541–549.

Walker, A. G.
1955 Connexions for parallel distributions in the large I. Quart. J. Math. Oxford (2) *6*, 301–308.
1958 II. Quart. J. Math. Oxford (2) *9*, 221–231.

Weinstein, A.
1977 Lectures on Symplectic Manifolds. Regional Conference Series in Math., vol. *29*. Providence, RI: Amer. Math. Soc.

Whitehead, J. H. C.
1952 Elie Joseph Cartan 1869–1951. Obituary Notices of Fellows of the Royal Society *8*, 71–95. Works I, 331–355.

Whitney, H.
1935 Sphere spaces. Proc. Nat. Acad. Sci. USA. *21*, 462–468.
1937 Topological properties of differentiable manifolds. Bull. Amer. Math. Soc. *43*, 785–805.
1941 On the topology of differentiable manifolds. In: Lectures in Topology, pp. 101–141. Ann Arbor, MI: Univ. of Michigan.

Willmore, T. J.
1956a Parallel distributions on manifolds. Proc. London Math. Soc. (3) *6*, 191–204.
1956b Connexions for systems of parallel distributions. Quart. J. Math. Oxford (2) *7*, 269–276.
1957 Systems of parallel distributions. J. London Math. Soc. *32*, 153–156.

Winkelnkemper, H. E.
1982 The graph of a foliation. Ann. Global Anal. Geom.

Wolf, J. A.
1967 Spaces of Constant Curvature. New York: McGraw-Hill.

Wood, J.
1969 Foliations on 3-manifolds. Ann. of Math (2) *89*, 336–358.

Zieschang, H.
1965 Proc. IV All-Union Topology Conference, Tashkent 1963.

Zimmer, R. J.
1980 Strong rigidity for ergodic actions of semisimple Lie groups. Ann. of Math. (2) *112*, 511–529.
1983 Curvature of leaves in amenable foliations. Amer. J. Math.
1982 Ergodic theory, semisimple Lie groups, and foliations by manifolds of negative curvature. Inst. Hautes Études Sci. Publ. Math. *55*, 37–62.

Supplementary Bibliography

At various places in the text, indications are given of applications of foliations. For example, thermodynamics and control theory are discussed in § I.6 and rigid motions in general relativity are discussed in § IV.4. The purpose of this supplementary bibliography is to indicate briefly a few more such applications.

In general relativity, a 3-dimensional space-like foliation gives a meaning to simultaneity, at least locally. Moreover, a space-like submanifold such that the trace of the second fundamental form vanishes usually gives a local maximum for volume. In the following reference, a foliation such that all the leaves are maximal submanifolds is used in numerical studies of the collision of black holes:

L. Smarr, A. Čadež, B. DeWitt, and K. Eppley, Collision of two black holes: Theoretical framework. Phys. Rev. D *14* (1976). 2443–2452.

The existence of maximal space-like foliations is studied in the following papers:

M. Cantor, A. Fischer, J. Marsden, N. O'Murchadha, and J. York, The existence of maximal slicings in asymptotically flat space-times. Commun. Math. Phys. *49* (1976), 187–190.

F. Estabrook, H. Wahlquist, S. Christensen, B. DeWitt, L. Smarr, and E. Tsiang, Maximally slicing a black hole. Phys. Rev. D *7* (1973), 2814–2817.

B. Reinhart, Maximal foliations of extended Schwarzschild space, J. Math. Phys. *14* (1973), 719.

Space-like foliations are also useful in studying the initial value problem for Einstein's equations. In the following papers, this problem is studied for data given on a maximal submanifold, or more generally on a submanifold of constant mean curvature:

Y. Choquet-Bruhat, Problème des contraintes sur une variété compacte. C. R. Acad. Sci. Paris Sér. A–B *274* (1972), A682–A684.

N. O'Murchadha and J. York, Initial value problem of general relativity. I. General formulation and physical interpretation. Phys. Rev. D *10* (1974), 428–436.

J. York, Conformally invariant orthogonal decomposition of symmetric tensors on Riemannian manifolds and the initial value problem of general relativity. J. Math. Phys. *14* (1973), 456–464.

A spatially closed universe does not in general admit maximal foliations, and indeed the existence of a single maximal space-like surface implies an expanding-contracting universe, as is shown in:

D. Brill and F. Flaherty, Isolated maximal surfaces in spacetime. Commun. Math. Phys. *50* (1976), 157–165.

Some other applications of foliations to relativity theory are given in:

J. Marsden and F. Tipler, Maximal hypersurfaces and foliations of constant mean curvature in general relativity. Phys. Rep. *66* (1980), 109–139.

D. Szafron and C. Collins, A new approach to inhomogeneous cosmologies: Intrinsic symmetries. II. Conformally flat slices and invariant classification. J. Math. Phys. *20* (1979), 2354–2361.

Another physical application is in the theory of singular constraints in quantum field theory, as studied in:

A. Lichnerowicz, Les variétés de Poisson et leurs algèbres de Lie associées. J. Differential Geom. *12* (1977), 253–300.

The rank of a manifold is the maximal number of independent commuting vector fields that it admits. The foliation defined by such a family is used in determining the rank by:

E. Lima, Commuting vector fields on S^3. Annals of Math. (2) *81* (1965), 70–81.

H. Rosenberg, The rank of $S^2 \times S^1$. Amer. J. Math. *87* (1965), 11–24.

Given a Riemannian immersion, the family of planes on which the relative curvature operator vanishes forms an integrable plane field. The global properties of the corresponding foliation are used to study the immersion in:

B. O'Neill, Isometric immersion of flat Riemannian manifolds in euclidean space, Michigan Math. J. *9* (1962), 199–205.

B. O'Neill and E. Stiel, Isometric immersions of constant curvature manifolds. Michigan Math. J. *10* (1963), 335–339.

Index of Terminology

Index of Symbols

Ergebnisse der Mathematik und ihrer Grenzgebiete

A Series of Modern Surveys in Mathematics

1. Bachmann: Transfinite Zahlen
2. Miranda: Partial Differential Equations of Elliptic Type
5. Dieudonné: La Géométrie des Groupes Classiques
7. Ostmann: Additive Zahlentheorie. 1. Teil: Allgemeine Untersuchungen
8. Wittich: Neuere Untersuchungen über eindeutige analytische Funktionen
11. Ostmann: Additive Zahlentheorie. 2. Teil: Spezielle Zahlenmengen
13. Segre: Some Properties of Differentiable Varieties and Transformations
14. Coxeter/Moser: Generators and Relations for Discrete Groups
15. Zeller/Beekmann: Theorie der Limitierungsverfahren
19. Boas/Buck: Polynomial Expansions of Analytic Functions
20. Bruck: A Survey of Binary Systems
21. Day: Normed Linear Spaces
25. Sikorski: Boolean Algebras
26. Künzi: Quasikonforme Abbildungen
27. Schatten: Norm Ideals of Completely Continuous Operators
30. Beckenbach/Bellman: Inequalities
31. Wolfowitz: Coding Theorems of Information Theory
32. Constantinescu/Cornea: Ideale Ränder Riemannscher Flächen
34. Mumford/Fogarty: Geometric Invariant Theory
35. Gabriel/Zisman: Calculus of Fractions and Homotopy Theory
36. Putnam: Commutation Properties of Hilbert Space Operators and Related Topics
37. Neumann: Varieties of Groups
38. Boas: Integrability Theorems for Trigonometric Transforms
39. Szökefalvi-Nagy: Spektraldarstellung linearer Transformationen des Hilbertschen Raumes
40. Seligman: Modular Lie Algebras
41. Deuring: Algebren
42. Schütte: Vollständige Systeme modaler und intuitionistischer Logik
44. Dembowski: Finite Geometries
45. Linnik: Ergodic Properties of Algebraic Fields
46. Krull: Idealtheorie
47. Nachbin: Topology on Spaces of Holomorphic Mappings
48. A. Ionescu Tulcea/C. Ionescu Tulcea: Topics in the Theory of Lifting
49. Hayes/Pauc: Derivation and Martingales
50. Kahane: Séries de Fourier absolument convergentes
51. Behnke/Thullen: Theorie der Funktionen mehrerer komplexer Veränderlichen
52. Wilf: Finite Sections of Some Classical Inequalities
53. Ramis: Sous-ensembles analytiques d'une variété banachique complexe
54. Busemann: Recent Synthetic Differential Geometry
55. Walter: Differential and Integral Inequalities
56. Monna: Analyse non-archimédienne
57. Alfsen: Compact Convex Sets and Boundary Integrals
58. Greco/Salmon: Topics in m-adic Topologies
59. López de Medrano: Involutions on Manifolds
60. Sakai: C^*-Algebras and W^*-Algebras
61. Zariski: Algebraic Surfaces

62. Robinson: Finiteness Conditions and Generalized Soluble Groups, Part 1
63. Robinson: Finiteness Conditions and Generalized Soluble Groups, Part 2
64. Hakim: Topos annelés et schémas relatifs
65. Browder: Surgery on Simply-Connected Manifolds
66. Pietsch: Nuclear Locally Convex Spaces
67. Dellacherie: Capacités et processus stochastiques
68. Raghunathan: Discrete Subgroups of Lie Groups
70. Kobayashi: Transformation Groups in Differential Geometry
71. Tougeron: Idéaux de fonctions différentiables
72. Gihman/Skorohod: Stochastic Differential Equations
73. Milnor/Husemoller: Symmetric Bilinear Forms
74. Fossum: The Divisor Class Group of a Krull Domain
75. Springer: Jordan Algebras and Algebraic Groups
76. Wehrfritz: Infinite Linear Groups
77. Radjavi/Rosenthal: Invariant Subspaces
78. Bognár: Indefinite Inner Product Spaces
79. Skorohod: Integration in Hilbert Space
80. Bonsall/Duncan: Complete Normed Algebras
81. Crossley/Nerode: Combinatorial Functors
82. Petrov: Sums of Independent Random Variables
83. Walker: The Stone-Čech Compactification
84. Wells/Williams: Embeddings and Extensions in Analysis
85. Hsiang: Cohomology Theory of Topological Transformation Groups
86. Olevskii: Fourier Series with Respect to General Orthogonal Systems
87. Berg/Forst: Potential Theory on Locally Compact Abelian Groups
88. Weil: Elliptic Functions according to Eisenstein and Kronecker
89. Lyndon/Schupp: Combinatorial Group Theory
90. Edwards/Gaudry: Littlewood-Paley and Multiplier Theory
91. Gunning: Riemann Surfaces and Generalized Theta Functions
92. Lindenstrauss/Tzafriri: Classical Banach Spaces I
93. Besse: Manifolds all of whose Geodesics are Closed
94. Heyer: Probability Measures on Locally Compact Groups
95. Adian: The Burnside Problem and Identities in Groups
96. Halmos/Sunder: Bounded Integral Operators on L^2 Spaces
97. Lindenstrauss/Tzafriri: Classical Banach Spaces II
98. Priess: Angeordnete Strukturen
99. Reinhart: Differential Geometry of Foliations
100. Petersson: Modulfunktionen und quadratische Formen

Springer-Verlag
Berlin Heidelberg New York Tokyo